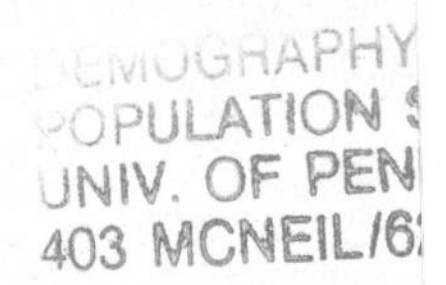

# BRITAIN'S POPULATION

Many current issues of social concern relate to the changing structure, organization and developments of Britain's population. The 'family' is no longer typified by two married parents and 2.4 children, but rather by a wider variety of social units – including a large number of single-parent families and re-formed families from previous marriages. Britain also has an ageing population. With less adults in work and more dependent pensioners, who will care for the elderly in the years ahead?

*Britain's Population* addresses these and other issues relating to the demographic characteristics of British society. Many of the contemporary features of the population relate to change in the past – particularly the ups and downs in attitudes to marriage and family formation. The history of these trends is considered, including the 'baby boom' of the 1960s when three million children were added to the population within the space of ten years. Jackson argues that the impact of this bulge generation can still be identified and will become of increasing importance when the generation reaches retirement age. Current trends in fertility are influenced by the changing structure of the labour market and by the delay in marriage and childbearing to later in life. The 1990s has been the era of the 'double income no kids yet' partners and the thirty-something mother.

Many of these issues have been included on the political agenda because of their impact on public policy and welfare spending; in this book Stephen Jackson highlights how the plight of single mothers, the problem of funding pensioners, and the future of the welfare state, all depend on demographic trends in society. *Britain's Population* will be invaluable to both students of demography and population studies and to all those with a general interest in contemporary British society.

**Stephen Jackson**, a Geographer by training, is Assistant Provost, with responsibility for teaching and learning, at Liverpool John Moores University.

# BRITAIN'S POPULATION

## Demographic issues in contemporary society

*Stephen Jackson*

London and New York

First published 1998
by Routledge
11 New Fetter Lane, London EC4P 4EE

Simultaneously published in the USA and Canada
by Routledge
29 West 35th Street, New York, NY 10001



Typeset in Garamond by
Florencetype Ltd, Stoodleigh, Devon
Printed and bound in Great Britain by
Creative Print and Design (Wales), Ebbw Vale



*British Library Cataloguing in Publication Data*
A catalogue record for this book is available
from the British Library.

*Library of Congress Cataloging-in-Publication Data*
Jackson, Stephen
Britain's population: demographic issues in contemporary
society/Stephen Jackson
p. cm.
Includes bibliographical references and index.
(cloth: alk. paper)
(pbk.: alk. paper)
1. Great Britain—Population. I. Title.
HB3583.J33 1998 97–35291
304.6'0941—dc21 CIP

ISBN 0–415–07075–9 (hbk)
ISBN 0–415–07076–7 (pbk)

# CONTENTS

# FIGURES

# TABLES



## Appendix tables

# PREFACE

Over the past twenty years there have been a number of significant changes in the structure and characteristics of Britain's population. A decline in the number of marriages, an increase in divorce, more children born outside marriage, more single-parent families and a trend towards having children later in the life-cycle. For some this represents a fundamental change in the structure of society – the decline of the family and with it all the moral values associated with stable and secure communities. The purpose of this book is to explore the facts behind these and other issues, to examine the underlying trends in Britain's population and to relate the changes to contemporary social and economic developments.

The study of population does not fit neatly into the curriculum of higher education. It forms a part of a number of subject disciplines within the social sciences as well as having relevance for a wide variety of vocational subjects such as Business Studies, Health Care and Social Policy. The starting point for this study has been an interest in the population geography of Britain, both in the present and the past. However, this is not essentially a geographical study. It attempts to begin with an analysis of demographic issues and then relate the study of population to the broader context of British society. The approach brings together an explanation of the principal sources for study, methods of analysis and familiar topics of demography, with an overview of current debates about population issues. As a consequence the material referred to not only covers the standard statistical sources – largely produced by the Office for National Statistics, but also newspaper articles and other ephemeral publications. It is intended to show that an understanding of demography is relevant to the explanation of current social issues and that debates about matters of current concern need to be based on reliable statistical evidence.

The title of the book is *Britain's* population and this is the principal geographical focus of the study. There is a tendency to use the terms Britain, England and Wales, and United Kingdom almost interchangeably – without always acknowledging the precise geographical territories referred to. Britain comprises England, Wales and Scotland, but not Northern Ireland.

Much of the detailed statistical information is available for the component parts of the United Kingdom and it is possible to aggregate information to the appropriate level. However, some categories of information are only available at individual levels. For example, information on immigration is normally only presented for the United Kingdom. Where geographical areas, other than Britain, are referred to in the text, or in figures and tables, these are specifically identified.

Much of the statistical information used in the book comes from the publications of the Office for National Statistics. In particular, extensive use has been made of the material published every quarter in *Population Trends*, which contains a comprehensive range of statistical data on the key aspects of population change in Britain. Other important published sources include *Regional Trends*, *Living in Britain* (General Household Survey), *Key Population and Vital Statistics* and the reports from the 1991 Census. One of the major difficulties of writing a book on population is the fact that the Census is only conducted once every ten years. The optimum time to publish is two to three years after the Census when the new material is fully available. Unfortunately that window of opportunity has not been taken advantage of with this book, and material derived from the Census refers to conditions in Britain some years ago now: in the early 1990s. However, the book is mainly about the 1990s as a period in time and the contemporary focus is set in a context of recent changes. The dated nature of the census material is to an extent offset by reference to more contemporary material from newspapers and other sources.

## Acknowledgements

The author would like to thank Phil Cubbin and Elaine Hodkinson for their kind assistance in the production of Figures 6.3, 6.4, 6.5 and 6.6. Also thanks to the following for permission to reproduce copyright material: Office for National Statistics (Appendix tables A–J), and British Railways Board (Figure 6.7).

# 1
# INTRODUCTION

Information about Britain's population is essential for an understanding of the major social, economic and political issues of the 1990s. Basic facts about the structure of population by age-group and sex, the distribution of people within the country and particularly the underlying trends in population growth and changing patterns of demographic behaviour, have a direct influence on policy formation and decision-making by central and local government.

For example, Britain has an ageing population. There is a general need to cater for a long-term increase in the numbers in the post-retirement age-groups. Future plans for health-care provision must take account of this age distribution as it will have a significant impact on the demands placed on the National Health Service and on the allocation of resources. This shift in age structure also raises issues about a possible increase in the burden of dependency in future years. Too few people will remain in work to generate the taxes and superannuation payments to cover increasingly large outgoings on retirement pensions and funding for basic services. Part of the problem relates to general assumptions about the role of the elderly in society. An older population may not necessarily mean a less healthy population and there are many ways in which the elderly may play a very positive role in the future. But it is necessary to plan ahead for these changes and ensure that adequate provisions are made.

Similarly labour recruitment and youth training policies are affected by the sharp downturn in birth rates that occurred during the 1970s. Progressively fewer school leavers have entered the labour market over the past five years, at a time when the country needs an increasing number of well-trained and technically advanced young people to support current trends in economic development and to keep pace with the levels of technical and professional skill in other European countries. Other groups, such as married women with children, have been encouraged back into work to make up for the shortfall. But this in itself poses problems about providing adequate care and nursery education for young children. It has also required the relaxation of the rigidities of the working day to allow for more flexible working patterns.

It is not only in these more obvious areas of development that population issues are important. There are many areas of pressing concern for which basic demographic data are required. Policies for inner-city regeneration, for example, must be based on a detailed knowledge of the 'localities' involved (age structure, social characteristics, employment structure, ethnic groupings), as well as an understanding of more recent patterns of population movement resulting in the sustained loss of numbers from the major metropolitan centres. The problems of inner cities are not so much to do with places as with people and it is knowing who these people are, their needs and aspirations, that is the important first step towards planning for redevelopment.

This book will attempt to provide some of this background demographic information by giving a broad overview of the principal characteristics of Britain's population and by discussing the likely future trends. There is no shortage of statistical material for attempting this task. A number of government departments are responsible for the systematic collection of information about many aspects of our daily lives. The decadal population census and the civil registration of births, deaths and marriages are perhaps the most comprehensive sources, but in addition there are many other types of information dealing with general social conditions, regional variations, employment, immigration, housing, health, etc., which provide a vast array of detailed material. The Office for National Statistics (formerly the Office of Population Censuses and Surveys) is the primary government agency responsible for collecting and publishing demographic data, but many other departments also collect data which are related to population issues. Some care is required in interpreting this published statistical material. Often it is necessary to question not only the reasons why the information has been collected, but also the assumptions and definitions that lie behind its organization and presentation. For example, unemployment figures in Britain have been significantly influenced by numerous changes in the definition of unemployment status during the 1980s and 1990s.

Part of the problem of interpreting demographic data relates to the use of standardized methods of statistical presentation. A knowledge of the more commonly used techniques is invaluable in understanding what is being discussed. The significance of measures such as the 'standardized mortality ratio' or the 'total period fertility rate' may not at first sight be self-evident. But their calculation allows for a much greater degree of precision in the assessment of demographic variation, both between different areas and different time periods. In addition, the study of demography invariably requires the classification of the population into meaningful groups (by economic activity, social status, ethnic status, etc.). The methodologies used for achieving such categorization have varied over time and again it is helpful to know exactly what is meant when reference is made to 'ethnic minorities' or to the members of 'social class IV'. The assumptions made in classification may well prejudice the interpretation of population statistics, particularly when investigating topics such as the variation in fertility and mortality between different places and between different groups in society.

An understanding of contemporary population patterns must also rely on some knowledge of historical demography. Changes in the past have a continuing influence on population dynamics, as for example with the variations in the birth rate in the years since the Second World War. These variations have created generations of different sizes within Britain's population structure which will be of significance for many years to come. An explanation as to why these changes occurred is important to our appreciation of the relationship between population and other social and economic factors, and is of particular relevance to the forecasting of future population changes.

Studying population change over time emphasizes the importance of the key demographic determinants of fertility, mortality and migration. Each requires separate treatment although clearly they interrelate with each other. Variations in fertility are related to social, economic and behavioural influences, some of which can be readily identified (such as disposable income), but others are very much more difficult to define or assess (such as changes in attitudes towards marriage and family formation). Prevailing trends in mortality are closely bound up with environmental and medical factors. Britain, in line with other European countries, has a low rate of overall mortality, but there is potential for improvement in specific areas, such as neonatal and infant deaths and also adult deaths in the age-range 45–65. Additionally, there are marked regional variations in mortality which have been persistent over many years. An appreciation of the reasons for these may have some bearing on policies for health-care provision and resource allocation.

Migration to and from Britain has, in recent years, been on a relatively small scale in comparison to overall population size. But immigration in particular has been a politically sensitive issue. This is partly because of changes in the policies towards immigration control and partly because of the perceived longer-term impact of migrant communities on British towns and cities. The importance of 'ethnic minorities' as part of a multicultural society raises many issues about social policy (housing, education, local politics, etc.), as well as focusing attention on the process of cultural assimilation. Future trends in population movement are likely to be more closely focused on the European Union rather than on the countries of the British Commonwealth, particularly if the membership is expanded to include the countries of Eastern Europe.

Analysing population characteristics at national level invariably involves making broad generalizations. Regional patterns of demographic change provide a clearer indication of the impact of social and economic developments over the last twenty years. This is particularly true in respect of inter-regional movements. The decline of the more traditional manufacturing economies in the North and West of Britain, in association with the advance of an increasingly service-orientated South East economy, has led to a redistribution of population and to the growth of the areas on the edges of the expanding metropolitan centre (South West, East Anglia). A detailed study of population change within the South East region highlights the fact that there are a number of different forces in operation. It is not just a question of the attraction of economic opportunity and the concentration of wealth within the region sucking people towards the centre, but also there are the negative effects of a high cost of living and the expense of housing at the centre, forcing people towards the edges of the region and creating a backwash effect on other areas.

Clearly, economic factors are of paramount importance in explaining regional problems and provide the driving force behind population developments. People are both consumers and producers within the British economy, and changes in population structure and distribution have a profound effect on regional economic performance and on the geographical unevenness of development. Of specific concern is the need to match the skills, capabilities and distribution of the labour force to the specific requirements of industry, commerce, administration and the professions. The education and training of young people will be a critical concern over the next ten years as the British economy becomes more closely aligned with the rest of Europe. At present the relatively low rates of achievement in training and higher education leave the country unable to match the levels of productivity of competitors like France and Germany.

However, it is not sufficient simply to view people as individuals within the economic framework of the country. There are many aspects of life which are more closely tied in with social characteristics. Education, housing, health care and opportunities for employment can all be related to underlying demographic variations between different social groups. Some issues are of direct concern, such as the variations in infant mortality between the extremes of the social range. Others are less easily identified, such as attitudes towards the family. But nonetheless it is clear that social factors remain a major consideration in the evaluation of current population trends. All of the major issues in the 1990s have a clear social dimension and many of these issues are increasingly becoming incorporated into the political agenda. The role of women, the care of the elderly and the disabled, the organization of family life, and the values associated with a secure family upbringing have all received a renewed significance in recent years.

Many of these issues will require a fuller examination and assessment over the coming years because of their direct impact on the organization of life in Britain. This is not to say that the country needs a 'population policy' geared towards managing fertility and mortality, but rather that demographic factors should be given adequate consideration in the broader context of economic and social planning. Decisions about the provision of health care, the availability of nursery education, the assistance for single-parent families, the resourcing of education, the regeneration of inner cities, etc. must first and foremost focus on the people involved. Planning for the future requires details about the present and an understanding of the factors that determine population change.

# 2

# SOURCES OF POPULATION DATA

The organization and management of an advanced industrial society like Britain requires the systematic collection and processing of a mass of information by central government. Most aspects of daily life are monitored and recorded through established and sophisticated mechanisms for data acquisition and analysis, and the details obtained form an essential element in the day-to-day administration of the country. Some of this material is the subject of general concern and contemporary political discussion. Statistics such as the retail price index, the rate of inflation or the level of unemployment are very familiar and are taken as measures of current economic and social trends, as well as indicators of the success or failure of government policy. They are regularly reported and commented on in the media. Such 'headline' figures are clearly significant because of the way in which they are interpreted, but there are many contemporary issues for which precise statistical information is essential to gain a clear understanding of underlying developments. The levels and changing patterns of crime, the educational attainments of school leavers, the performance of the National Health Service, the availability and price of housing, etc., can only be assessed effectively by reference to up-to-date and accurate figures, although the statistics provided are often subjected to varying interpretations.

As a consequence, central government produces a massive amount of statistical data about Britain's population. Institutions, businesses, government agencies, services and the general public provide this information in a variety of different forms on a regular basis. Details about the social and demographic characteristics of the population are collected systematically through the use of surveys and the registration of vital events. In addition, there are many indirect ways in which information about the general public is recorded. For example, when people change address and register with a different family doctor, the information is recorded by the National Health Service Central Register and forms part of a data set used for monitoring internal migration.

Almost every time a form is filled in or details of name and address are provided, information about individuals is gathered by government or other institutions or commercial organizations. The confidentiality of this information is protected by legislation which effectively restricts its use to pre-defined purposes. Hence the situation has not yet been reached in this country whereby all aspects of the population can be systematically monitored through the linkage and cross-referencing of data files on different subjects, although statistics derived from some data bases may be useful for illustrating individual topics, for example information on the purchase of cars from the Driver Vehicle Licensing Centre. At present the sources most readily available for the study of population are largely those collected by central government for the express purpose of monitoring current demographic and social trends.

Information about changes in the size and structure of Britain's population is available from three principal sources: the census, which provides a cross-section of the population and contains a wealth of detail about individuals and households; civil registration, which records the dynamics of population change (births, deaths and marriages); and statistics on migration, which measure the flow of people into and out of the country as well as detailing the movement of individuals between regions and local areas.

## The census

Information on the principal characteristics of Britain's population is systematically collected every ten years by the census of population. The Office for National Statistics (ONS) is responsible for conducting the survey in England and Wales. In Scotland it is handled by the General Register Office. The census provides complete coverage of contemporary demographic and social patterns and allows for the detailed investigation of population at a variety of scales (local, regional, national). Its main value is its comprehensiveness. Every household is required by law to complete a census form and enumerators make every effort to ensure that all individuals are included. The data are collected and processed following proven and well-regulated procedures which guarantee the uniformity of detail and which allow direct comparisons to be drawn between different areas. In addition, although there have been changes in the format and content of census forms in the past, there is a high degree of continuity of basic statistical detail which provides for the study of longer-term changes between successive censuses.

The primary purpose of the census is to provide information for decision-making and planning by central and local government, health authorities, business, service industries and the professions. Accurate information is essential for the allocation of basic resources, for the provision of services and for the determination of priorities. Many areas of responsibility such as health care, housing, employment, transport, education and welfare support can only be effectively managed with a data base of information which identifies needs and which provides a reliable basis for policy formation and implementation. Others make use of information from the census in order to analyse various facets of Britain's contemporary population and society. Interests range from academic research into patterns of population movement or variations in the basic quality of life between different localities, to commercial assessments of the social and demographic characteristics of small areas for the purposes of market analysis or defining of target populations.

One of the basic principles behind the design of the census is that only information for which there is an identifiable need should be collected. The ONS has to balance the desire to obtain as much detail as possible against the willingness of individuals to respond to the survey. The census for 1991 included twenty-four questions for each private household. The information collected falls into two principal categories: details about individuals (age, sex, occupation, educational qualifications, country of birth); and details about life-styles and living conditions (household structure, type of accommodation, tenure, amenities). The responses to the questions provide a matrix of census variables which can be cross-tabulated to provide a vast range of socio-demographic indicators.

New developments for the 1991 census included questions on ethnicity and illness. The collection of detailed information on ethnic origin has been a sensitive issue for many years. The 1971 census included questions on country of birth and birthplace of parents in an attempt to identify the distribution of recent migrant populations. In 1981 a specific question on ethnic origin was dropped after unsuccessful pre-census tests in the London borough of Harringay (Thatcher 1984). However, the 1991 census asked respondents to identify their ethnic origins by category (White, Black Caribbean, Black African, Black Other, Indian, Pakistani, Bangladeshi, Chinese, Other). Although reservations remain about the definition of each category, very few people raised objections to the question in the 1989 pre-census test (White 1990). The availability of detailed statistics on ethnicity provides a clearer picture of the social and demographic characteristics of individual groups and by cross-referencing this information with other social indicators (housing, employment, etc.) it is possible to identify the extent of racial disadvantage and the existence of special needs of ethnic minorities.

Limitations on space only allowed for a single question on health which attempted to cover a range of different issues. A question on long-term illness was designed to give a better indication of relative need for health-care provision and the extent of support from carers within households. The question asked: 'Does the person have any long-term illness, health problem or handicap which limits his/her daily activities or the work he/she can do?' Respondents were asked to include problems which were due to old age but were not asked to specify any detail. As a consequence there was undoubtedly some variation in the interpretation of the question and the reliability of the responses may be dubious. Nevertheless, this is the first time that details of this kind have been collected from every member of the population and the information allows for the identification of health-care requirements on a much more localized basis than has previously been possible.

The information gathered by the census is made available in a number of different forms covering a variety of topics at different geographical scales. Publication of information in the form of reports takes time and the results are produced periodically over a three-year period after the date of the census. For 1991 the ONS has made the more detailed information available in the form of computer-readable output, either for use on-line from major computing centres or for inputting into micro-computer systems. The essential building blocks for this information are the local base statistics, which cover every census topic for all areas down to ward level (20,000 tabulated counts). Local base statistics can be used to create reports and statistical abstracts by aggregating units into larger blocks and by selecting and cross-tabulating variables. Information may also be derived for smaller areas. The SAS (small area statistics) provide information for every enumeration district (130,000 in Britain) for up to 9,000 tabulated counts from the local base statistics. Enumeration districts are based on postcode areas in Scotland which allow the SAS to be cross-referenced with other data available at this scale. The ONS has also released information by postcode sector for England and Wales.

The SAS provide a highly flexible and comprehensive data base of statistical material which can be tailored to meet the needs of a wide variety of different users. Increasingly sophisticated computer software, particularly the development of Geographic Information Systems, enables finely detailed analyses of individual areas to be produced from the data. Cross-referencing census data with other information available for small areas has led to the development of accurate profiles of the residents of specific localities and to the identification of a mix of variables which denote different life-styles. Without infringing the confidentiality of the census, these procedures have greatly enhanced the usefulness of census material so that instead of dealing with generalized patterns of social indicators it is now possible to identify local characteristics with a much greater degree of resolution (Table 3.7).

The principal dimensions of the national population are contained in a series of reports consisting of tables dealing with particular topics. These cover subjects such as household and family composition, economic activity, migration, travel to work, ethnicity, and illness. Most of the tables in these reports refer to national areas – Britain, Scotland, England and Wales – or to standard regions, but as the information is built up from the local base statistics it is possible to obtain information for smaller areas. The ONS will prepare specialized sets of information including selected census variables and cross-tabulations, on request. Another development with the 1991 census has been the availability of samples of anonymized records (census forms with the names deleted). Two different samples have been constructed covering individuals and households. Their value is mainly for social and demographic research, and the information has been produced at the request of the Economic and Social Research Council.

The range and detail of information available from the 1991 census represents a significant step forward in terms of what is known about Britain's population. Attempts were made to achieve total coverage of all Britain's inhabitants. Prior to the day of the census, enumerators conducted an advance survey of properties in each district to obtain an assessment of the total number of households and, following the initial collection of census forms, they returned to houses where no response had been given. In addition to covering individual households and communal institutions, 2,700 people 'sleeping out' on census night were located and recorded. However, the information collected still contained a significant undercount. Approximately one million people were missed by the census for various reasons and the response rate in 1991 was not as high as it had been in 1981 (Wormald 1991). The greatest difficulties were encountered in inner-city areas where, amongst other problems, there were fears that information provided for the census might be used to monitor 'poll tax' registrations and track down defaulters (*Guardian* 1991a). These shortcomings make it difficult to obtain a precise picture of the characteristics of some of these more problematic localities. The central areas of British cities have witnessed significant changes over recent years, including a sustained loss of population through out-migration. They are also the places which show the greatest ethnic diversity and traditionally contain many of the more severe problems of social deprivation. For both central government and local authorities these communities are likely to be the focus of much continuing attention in the future and they will require accurate and detailed information in order to manage their redevelopment.

The problem of ensuring comprehensive coverage together with the expense of collecting and processing the data (£135 million for the 1991 census) have raised the issue of whether periodic large-scale surveys are the most effective way of monitoring the nation's population. By the time much of the census data is published it is already out of date and the ten-year gap between surveys makes it difficult to maintain an accurate account of changes within the population – particularly for local areas. A lot of the detailed information provided by individuals for the census is already available to a variety of different central government agencies or other organizations. For example, information on date of birth is collected through civil registration and is available on medical records; details of employment status are held by the Inland Revenue; the Department of Social Security has information about all those claiming benefit; educational qualifications are held by examinations boards and institutions of higher education; car ownership details are kept by the Driver Vehicle Licensing Centre at Swansea, etc. In addition, much of the information that is collected about living conditions is repeating, at a more detailed level, surveys which are conducted on a regular basis using sampling techniques (such as the *General Household Survey*, and the *Labour Force Survey* – see pages 18–20).

With an increasing amount of this detail about individuals and households being held on computer data bases, and with the enhanced capacity and capabilities of information technology systems, it is possible to develop a national population register which would bring together information from all of these different sources. A regularly maintained register would provide contemporary statistical material and do away with the need for decennial censuses. It could also replace other registers of information about individuals (e.g. electoral registers) and could greatly enhance the efficiency of central and local governments and health authorities. It would not be a new development, in that Britain has had a population register before. During the Second World War a national register was established for the purposes of rationing and state security. This register continued in use until the end of rationing in 1952 and was used as the basis for the National Health Service Central Register (Fox 1990).

However, the introduction of a population register would undoubtedly raise questions about civil liberties and about the security of personal information. Registers are associated with the introduction of identity cards, a subject about which the British public has traditionally been very sensitive. It would also be very expensive to set up a register in the first instance. In countries where population registers have been developed, long delays have been encountered in the production of information, and they rarely provide the depth and range of coverage of the British census (Thatcher 1984). These failings suggest that centralized record-keeping on individuals may not be introduced in the near future although it may be possible to develop a greater degree of cross-referencing between existing repositories of information. In the longer term it is likely that the tradition of census-taking in Britain will alter in response to the rapid growth of information processing, particularly in the commercial sector, which is creating increased demands for detailed and up-to-date information on individual areas. Accurate data are needed to achieve more precision in market research and to monitor changing trends. In addition, the move towards closer integration in the European Union will require the standardization of statistical information and the harmonization of census-taking across all member states.

## Civil registration

Despite any reservations there may be about the development of a national population register, the actual registration of vital events (births, marriages and deaths) is a well-established and familiar procedure. Registration is required by law and fulfils two basic functions: it provides an accurate statistical record of changes within the population and it provides confirmation of personal identity – legal proof of nationality and status which is required, for example, to obtain a passport or to provide evidence of age or marital condition.

Registration is organized at a local level by staff appointed and paid for by local authorities, but with responsibilities to the Registrar General. These offices handle approximately 1.5 million registrations annually (Nissel 1987). The information they process is fed through to the ONS and eventually housed on microfilm in the Family Records Centre in London where it can be accessed, for a fee, by the general public.

The information collected by the registration service goes beyond the simple recording of basic events. Additional details are asked for which are used for analytical purposes. These provide an important insight into changing trends in demographic behaviour and social attitudes. From the outset of civil registration in 1837 it was recognized that the regular gathering of statistical information about the population was essential to the identification and measurement of patterns of mortality and fertility. At a time when death rates were high, especially in urban areas, it was necessary to analyse the factors responsible to assist in the formation of social policy. Dr William Farr, the first statistician in the General Register Office, produced detailed statistics on the cause of death, which highlighted the significance of environmental conditions in mid-nineteenth century towns and which influenced the development of legislation on sanitary reform and building control. In more recent years, changes in attitudes towards family life and the relative decline of marriage have been identified from birth and marriage registrations (Whitehead 1987, Haskey 1990). The recording of place of birth of parents on birth certificates has allowed for some estimation of levels of fertility amongst different ethnic groups. In addition, the linkage of records of vital events to census information by the ONS *Longitudinal Study* (see page 21) provides a very detailed insight into the life-cycle of individual families.

The registration of births is required within forty-two days of the event by the parents or others with responsibility for the child. The information recorded not only includes details of the child (name, sex, date of birth) but also of the parents. The age and place of birth of both parents, the occupation of the father, the date of marriage and the number of previous births to the mother are noted. Information on the father is not required for illegitimate births although increasingly such births result from stable unions and are recorded jointly by both partners (Ermisch 1990). Still-births (children not born alive after twenty-eight weeks' gestation) are recorded in the same way with the addition of a statement from the doctor or midwife in attendance on the mother detailing the cause of the still-birth. Births are also recorded separately by the National Health Service to enable health visitors to make contact with parents. These listings contain additional information on, for example, birth weight and they are widely used by local registry offices to double-check the accuracy of birth returns.

The registration of death must take place within five days of the event and requires a certificate stating the cause of death from the medical practitioner attending the deceased in their last illness. The details recorded include date, place and cause of death, and the name, sex and age of the deceased. Information is also required about the occupation of men and single women. Father's occupation is recorded for children and husband's occupation for married women. With women making up nearly 44 per cent of the present labour force this provision is perhaps a little dated. Details of cause of death and occupation are of significance for the analysis of mortality by social and occupational groupings and for the monitoring of accidents and fatal diseases.

Marriages are recorded at the event (either civil or religious) and the certificates provide legal confirmation of a union. Information is provided about the status of each partner (single, divorced, widowed), their occupations and residence at the time of marriage. In addition, details are recorded for each of the fathers (name and occupation). Legal separation by divorce is not recognized as a vital event and is not recorded directly by the registration system. However, statistical information on divorce is available from the courts and is published by the ONS (Benjamin 1989).

The constant inflow of information from the registration system allows for the monitoring of population dynamics at national, regional and local levels and provides an indication of underlying population trends. This information (together with details on migration) is used to estimate the principal changes in population between census years and to produce population projections. The detail compiled from registration returns is published by the ONS in a series of reports and monitors (*Birth Statistics*, *Conceptions*, *Mortality Statistics*, *Cancer Statistics*) and also as statistical information on computer disc (*Annual Births and Deaths Extracts*). Much of this data is also reproduced in other quarterly and annual publications including *Population Trends*, *Regional Trends* and the *Annual Abstract of Statistics* (see Table 2.1). In addition to the regular publication of the statistics of births, deaths and marriages, more specialized reports are produced periodically which investigate various aspects of current trends and present a more consolidated view of longer-term changes. Included in this category are reports on mortality and geography, occupational mortality, period cohort birth order statistics and historical series for births, marriages and divorces.

The procedures involved in recording and transferring the information from initial registration to published sources have changed little since the introduction of civil registration. The recording of births, marriages and deaths is still done by making a handwritten entry in the local register and by writing out individual certificates. Every three months copies of the returns are sent in to the central office. The development of more advanced data storage and retrieval techniques is limited, although clearly the potential exists for the entry of information at local offices directly to a central data base. Running totals of births, deaths and marriages could then be maintained and up-to-date information could be accessed either centrally or from local offices. Such proposals again raise questions about expense and data protection and may be difficult to implement at present.

The legal purpose of registration is also very significant. The existence of a birth certificate is the individual's ultimate guarantee of citizenship and ensures access to the services of the state and the right to vote. Certificates are required in matters of property ownership, inheritance and taxation and are generally regarded to be authoritative and indisputable. The information is also invaluable for historical research. Many of those who actually consult the documents in the Family Records Centre are tracing individuals in order to piece together their own, or other people's, family histories.

## Statistics on migration

Civil registration provides an accurate record of births and deaths in Britain and by and large the trends identified follow a predictable pattern. Given knowledge of the age structure of the population and the prevailing levels of fertility and mortality, it is possible to make precise estimates of the future impact of natural change (the contribution to population change accounted for by the positive input of births and negative output of deaths). There is not the same reliable and comprehensive record of information on migration from which to estimate overall trends in population change. In recent years there has been very little natural growth in the national population. The movement of people, particularly at a local and regional level, has been the most important factor in determining the differences in patterns of population growth and decline, not just in terms of the redistribution of individuals but also in terms of the impact that migration can have on the age composition and fertility and mortality levels of individual localities.

In contrast to the centralized and meticulously organized procedures established for recording vital events in Britain, the collection of information about population movement is undeveloped and fairly arbitrary. Unlike some other European countries, there is no central register of population movements and no statutory requirement on individuals to record a change of address. The information that is available is derived from three principal sources:

1 the census which asks for details of address the previous year;
2 sample surveys which monitor relatively small groups of people or particular categories of migration;

3 various surrogate sources which record changes of address or new registrations.

The National Health Service Central Register is the most important organization in this third category. It records details of all changes in registration with family practitioners. Other organizations also collect details of change of address (banks, utilities, etc.) but this information usually only refers to householders and is not generally made available for research purposes.

Patterns of international migration to and from Britain have been of considerable interest in the recent past (see Chapter 6). The actual scale of movement has been very limited, with migration adding only 62,000 (0.1 per cent) to the population of the UK in 1994 (*Population Trends* 87, table 17). This positive net inflow has only occurred over the past few years; before 1983 Britain tended to lose more people than it gained through migration. Of more relevance are the changes in areas of origin and destination of migrants. The inflow to Britain from non-European countries is tightly restricted by legislation on immigration. At the same time there has been a decline in the outflow of emigrants to more traditional destinations in Australia, New Zealand and Canada, where employment prospects dried up during the recession years of the early 1980s. By contrast there has been a growth in movement within the European Union.

The main source of information for these changing patterns of migration is the *International Passenger Survey*. This is a questionnaire survey carried out by the ONS on behalf of the Department of Trade and Industry. A stratified random sample of passengers arriving and leaving the UK is questioned at ports and airports to obtain information not just about migration but also more generally about international passenger movements, tourism and the estimated levels of expenditure by visitors. Approximately 180,000 interviews are conducted each year (Nissel 1987). Of these about one-third are incoming visitors and two-thirds outgoing. Only 2 per cent of those interviewed are classified as migrants (having stated their intention to remain in the UK, or abroad, for more than a year).

The information gathered by the survey is subject to considerable margins of error because of the relatively small sample size and because of the problems of achieving a truly representative sample. The statistics on international migration are published as an annual report by the ONS, which gives details of the country of last or intended future residence, sex, age, occupation, citizenship, marital status and route taken. Some details are also published quarterly in *Population Trends* and in the *Annual Abstract of Statistics*.

Estimates of net migrational change can be measured at different geographical scales within Britain. Short-distance moves are very common and, unless they involve the crossing of local or regional boundaries, they are not usually classified as migration. Flows between local areas and regions can be estimated by noting the population change between census years and subtracting the element of change accounted for by births and deaths. The residual element indicates the net contribution from migration which can be further subdivided into internal and international migration. The overall balance between migration gain and loss is only part of the picture. The calculations give little indication of the actual scale of inflow or outflow for a particular area or about the characteristics of those involved in the movements. Evidence from the census of 'usual address one year ago' allows for a very much more detailed analysis of individual population movements. The ONS publishes reports from the census which give cross-tabulations of population movements for all areas down to the level of districts. The information provided includes migrant's age, sex, type of household, economic position and employment status. Although this gives a very comprehensive overview of internal migration, the analysis is only possible once every ten years and it is based on information which may not cover all types of movement. Only those who have moved within the previous year are recorded and the assumption is made that individuals have only moved once. Patterns of population movement are constantly changing, particularly at the local level, and census data can only give a snapshot view of these patterns at one point in time. Other information about migration is collected on a more regular basis by the ONS, for sample populations, by the *General Household Survey*, The *Labour Force Survey* and the *Longitudinal Study* (see pages 18–21).

Since 1971, information on internal migration has been made available from the National Health Service Central Register (NHSCR). This contains records of all patients registered with the NHS (99 per cent of the population) and provides support for the local Family Health Services authorities by informing them of all changes to the register. Details of births, deaths, migration and enlistments in the Forces are recorded. It also tracks the movement of people between the Family Health Services authority areas by noting all re-registrations of individuals with family practitioners (Fox 1990).

The Family Health Services authority areas (formerly Family Practitioner Committee areas) correspond to the counties in England and Wales, to Metropolitan districts and to groupings of London boroughs. The information extracted by the NHSCR only includes details of the age and sex of patients re-registering and the areas they have moved from and to. Therefore it is not possible to use these data to study population movement at the same spatial scale or in the same detail as the census reports. Also there are some doubts about the comprehensiveness of the coverage. The identification of a move is dependent on patients re-registering with their local practitioner and not all groups do this consistently. Many will leave it until the services of a doctor are required and some, particularly young people, may move several times without re-registering (Devis 1984). However, comparisons between the NHSCR data and the census suggest a reasonable degree of consistency and confirm that the information is sufficiently reliable to indicate the principal trends in internal migration. The information is of greatest value for identifying changes in the rate and direction of movement over time.

Information from the NHSCR is used by the ONS to produce population estimates for individual areas between census years. It is also supplied directly to local authorities either as printed tables or in machine-readable form. The principal published source of the data is the ONS *Key Population and Vital Statistics for Local Health Authority Areas* (Series VS/PPI) which is produced on a regular basis. Information for counties also appears annually in *Regional Trends*.

Some indication of population movements over short distances within local authority or health areas can be obtained from electoral registers. Registers are compiled on an annual basis and cover each polling district within parliamentary constituencies. The details recorded for each household include the surname and first names of all those entitled to vote (aged 18 and over) or who will be entitled to vote during the period of the register. Information is also collected (but not published) for those aged 65 and over. Levels of population turnover can be estimated by taking account of recorded births and deaths – the residual figure provides an indication of net migration. The method lacks precision because electoral registers do not cover the whole population. Allowance has to be made for children and for those who fail to register. There are also problems of double-counting people who own more than one property or who change address. The updating of electoral registers offers the potential for the collection of more detailed information about individuals and households on an annual basis. Proposals for this form of yearly census have been considered in the past, but they have generally met with resistance from local authorities (Population Statistics Division OPCS 1980).

Records of population structure, vital events and migration form the basis of the study of population change, but in addition there is a vast amount of statistical information produced by central government agencies and other organizations which detail a wide variety of other demographic and social characteristics of the country's population. A comprehensive guide to the range of material, together with some indication of the detail contained, is provided by the *Guide to Official Statistics* published by the ONS. It would be impracticable to list all separate publications here, but a summary of the relevant agencies with details of the most useful publications follows.

## The Office for National Statistics

The ONS is the primary government agency responsible for gathering and disseminating information on Britain's population. It was founded in 1996 by the merger of the Office of Population Censuses and Surveys (OPCS) and the Central Statistical Office (CSO). OPCS itself had been established in 1970 as a result of the bringing together of the General Register Office (which managed the census and civil registration) with the Government Social Survey (responsible for sample surveys on various social issues). The association provided the opportunity for a greater degree of integration between different sources of statistics, and for the development of data-gathering procedures which were better coordinated. Many of the topics covered in the census are investigated in greater detail by the use of sample surveys which can be conducted far more cheaply and at more regular intervals. They also allow for the periodic investigation of topical issues and specialist subjects. The Government Social Survey became the Social Survey Division within OPCS and has remained as a separate section in the ONS.

The principal publications of the ONS are listed in Table 2.1 under different headings. ONS Monitors are information sheets which are produced at regular intervals and which make available recent statistical material and preliminary results. Much of this information appears in a consolidated form in later annual reports. Census reports cover all aspects of census results. Social survey reports list the output of the Social Survey Division, including major surveys such as the *General Household Survey* and the *Labour Force Survey*. Other publications cover a wide range of reports and commentaries on various social and demographic issues. The ONS also publishes the quarterly journal *Population Trends*, which contains regular tables on population totals, components of population change, live births, expectation of life, deaths, abortions, international migration, internal migration, marriage and divorce. Also included are articles on various topical issues, news of recent developments and details of ONS publications.

The work of the Social Survey Division provides an important insight into many aspects of daily life, highlighting changing attitudes and trends in social behaviour. The results are published in a number of reports. The *General Household Survey* (GHS) was introduced in 1971. It is a wide-ranging social study requested by a number of government departments. The principal results are published every year under the title *Living in Britain*. The survey is based on a continuous sample survey of approximately 12,500 private households selected from postcodes. The normal response rate is between 80 and 85 per cent. (Nissel 1987). Information is collected from every adult member of the household (aged 16 years and above). Six major topics are covered in each survey: housing (tenure, amenities, consumer durables, accommodation type and age); internal migration (time at present address, frequency of moves, birthplace); education (type of educational establishment, age on leaving, qualifications); employment (job description, unemployment); health (chronic and acute sickness, GP consultations, hospital treatment, use of other health-care facilities); and family information (details of family formation, marital history and cohabitation). In addition, other topics are included from time to time to broaden the scope of the

*Table 2.1* Principal publications of the Office for National Statistics

| *Title* | *Frequency* |
|---|---|
| Economy | |
| Economic Trends | Monthly |
| | |
| Health | |
| Abortion Statistics | Annual |
| Cancer Statistics | Annual |
| Health in England | Annual |
| Key Health Statistics from General Practice | Annual |
| Mortality Statistics | Annual |
| | |
| Labour market | |
| Labour Market Trends (Employment Gazette) | Monthly |
| Labour Force Survey | Quarterly |
| Annual Employment Survey | Annual |
| | |
| Population | |
| Population Trends | Quarterly |
| Birth Statistics | Annual |
| Electoral Statistics | Annual |
| International Migration | Annual |
| Key Population and Vital Statistics | Annual |
| National Population Projections (Monitor) | Annual |
| Population Estimates (Monitor) | Annual |
| | |
| Reference | |
| Monthly Digest of Statistics | Monthly |
| Annual Abstract of Statistics | Annual |
| Guide to Official Statistics | Annual |
| Key Data | Annual |
| | |
| Regional | |
| Regional Trends | Annual |
| Census County Reports | Deccenial |
| | |
| Society | |
| Family Spending (Family Expenditure Survey) | Annual |
| Housing in England | Annual |
| Living in Britain (General Household Survey) | Annual |
| Marriage and Divorce Statistics | Annual |
| Social Trends | Annual |

*Source: The Source, 97 Catalogue,* ONS (1997)

survey. Details have been collected on such issues as burglary, eyesight, hearing, dental health, smoking, drinking, leisure, heating, loans, sitting tenants, job mobility and absence from work. The results of the GHS provide a regular review of social change, but the relatively small sample size (0.025 per cent of the British population) means that information for individual groups (ethnic minorities, the elderly, etc.) may be unrepresentative. This problem is partly overcome by aggregating information over a number of successive surveys.

The *Family Expenditure Survey* (FES) is conducted by the ONS on behalf of the Department for Education and Employment. Its primary purpose is to collect information on household expenditure for the calculation of the retail price index, but much additional information is also recorded about the characteristics of households and the income of their members. The survey covers approximately 11,000 households on a continuous basis (response rate about 70 per cent) and the data are collected by interview and by the completion of expenditure 'diaries' during the fortnight following the interview. The detailed information obtained on living conditions (including ownership of a range of consumer durables), together with data on employment status and earnings for individual members of the household, provide a comprehensive series of indicators for studying variations in social conditions, considerably extending the information available from the census. However, the size of the survey effectively restricts the use of these data to the national or regional level. Annual reports are published by the Department for Education and Employment. Some of the material is published quarterly in *Labour Market Trends* (formerly *Employment Gazette*) which also includes articles discussing the main results of the survey. Data from the FES appear in various other government publications (*Regional Trends*, *Annual Abstract of Statistics*, *Economic Trends*, etc.).

The *Labour Force Survey* (LFS) was introduced in 1973 at the time Britain joined the European Economic Community. It is part of a community-wide survey of employment characteristics and is primarily intended for European Union policymaking on topics such as regional development and the distribution of the European Social Fund. The survey covers some 60,000 private households and since 1984 it has been conducted on an annual basis. Much of the information collected duplicates details available from the census or other surveys, but the LFS has a number of distinct advantages: it is based on a larger sample size than other household surveys; it is conducted more regularly than the census; and its results are presented in a format that is consistent with other European countries. For example, information on employment status provides an alternative source of information on unemployment which, unlike the standard statistics for Britain, has not been subjected to periodic redefinition in the past and can be used for comparative purposes with other countries. The principal categories of information for households are: size and composition, family structure, accommodation and tenure. For individuals they are: age, sex, marital status, nationality and country of birth, ethnic group, employment status (with details of present or previous jobs), second job details, situation one year ago, qualifications and current education. Quite apart from the value of the LFS for monitoring employment structure and change, much of the information is useful because of the limited coverage of certain topics in other surveys (for example, information on nationality and ethnic groupings). The annual reports are published by the ONS with preliminary results appearing, with commentary, in *Labour Market Trends*. The Department for Education and Employment retains the unpublished data.

Surveys provide information about people at given points in time. The ONS also monitors change through the *Longitudinal Study* (LS) which identifies the events occurring to one group of people over time. The LS was set up by taking a sample of all individuals born on each of four birth dates from the 1971 census. Records of the selected individuals were traced through the NHSCR and subsequently linked to returns in the 1981 and 1991 censuses. Since 1971 all those born on the selected dates (including immigrants registering with the NHS) have been added to the sample. Similarly, members of the sample who have died or emigrated have been removed. Linkage rates of individuals have been high. Ninety-seven per cent of the original sample in 1971 were traced to NHSCR records and 91 per cent of those still alive and in Britain were found in the 1981 census. Vital events occurring since 1971 have been added to the records, including information on cancer registrations. This has resulted in a data set of approximately 400,000 individuals for which profiles of life experiences can be studied and has proved to be remarkably useful for examining a number of different social, medical and demographic issues. Changes can be studied over the life-cycles of different groups of individuals, such as the varying employment status of women during family formation, the migratory moves of people at different times in their life, and the longer-term impact of unemployment and other poverty indicators on health and mortality. Reports produced so far have covered the topics of socio-demographic mortality differentials and social class and occupational mobility. A summary report of statistics covering the inter-censal period 1971–81 has also been published. In addition, information from the LS has been used for research into topics such as household structure and birth intervals, which has been reported in *Population Trends*.

## Other government departments

Although the ONS has primary responsibility for collecting and disseminating demographic and social data, many other government departments handle information which is relevant for research into population issues. All matters relating to employment and training are the responsibility of the Department for Education and Employment. It conducts the periodic Census of Employment (the most recent of which was held in 1987) which provides details of the population in employment. The Department's principal publication is *Labour Market Trends*, which is produced monthly and includes details on employment and unemployment together with articles discussing current trends. It also has overall responsibility for the publications of the Training, Enterprise and Education Division (formerly the Training Agency) which include the *Labour Market Quarterly Report* and other publications on skill requirements and training.

The Department for Education and Employment also produces *Statistics of Education*, with information on school leavers and examinations. The Home Office provides information on immigration and nationality, including *Control of Immigration: Statistics for the United Kingdom*, *Refugee Statistics*, and *Citizenship Statistics*. The Department of Social Security produces *Social Security Statistics* annually, as well as other publications detailing security benefits and personal social services. The Department of the Environment produces information on housing, households and inner cities.

This list is by no means exhaustive. Central government handles a vast amount of detailed statistical information, much of which is for specialized applications. The more important sources of data are now brought together by the ONS, which has taken over responsibility form the Central Statistical Office for collecting some of the information and for publishing consolidated volumes of statistics. Social and demographic data are included in the *Monthly Digest of Statistics*, *Annual Abstract of Statistics*, *Social Trends* and *Regional Trends*. Separate volumes of statistics are also produced for Scotland, Wales and Northern Ireland by the Scottish Office, Welsh Office and Northern Ireland departments. The whole system of information processing by government is managed and coordinated by the Government Statistical Service (GSS). This comprises the statistical divisions of all the major departments.

There are a number of other sources of data available at local and national level collected by non-government agencies and published in a variety of different reports and surveys. Organizations such as SHELTER, the Child Poverty Action Group and Age Concern collect information for the specific purpose of publicizing issues and lobbying government. In addition, many businesses, institutions, local authorities and the media make use of professional statistical consultants and data-gathering organizations (Gallup, National Opinion Polls, etc.) to collect information for particular projects or to investigate specific groups in the population. The problem with this type of information is its lack of comprehensiveness. It may be of value for studies at the local level or for widening investigations into particular topics, but from the perspective of population research the applications are generally limited.

The consequence of this is that almost all the information available for the systematic study of demographic issues is derived from government sources. The sheer amount of statistical information generated by the central government agencies is overwhelming. The GSS is by far the largest single provider of statistics in the UK and, although it has the responsibility for making the information it gathers available to anyone who requires it, its primary role is to service the needs of government and not specifically to advance the interests of research. Because of this it is necessary to exercise some care in the interpretation of published statistical material.

There are two important considerations: first the meaning and significance attached to certain types of data, and second the methods that are used to classify and define categories of information and the format of their presentation. In the first instance it is quite clear that certain types of demographic information can have a political relevance and may be used to support arguments in contemporary debate. A good example of this is the information published for levels of immigration and place of origin of migrants. Much attention was focused on these figures in the 1970s when immigration was a major political issue and there was much speculation about likely future trends in the size, composition and integration of ethnic minority populations. The availability of precise information was essential to place the debate firmly into context and counteract the rhetoric about the scale and impact of immigration.

Another much quoted example is the way in which figures for unemployment were revised and redefined during the 1980s. In the period 1979–89 there were no less than nineteen separate changes in the definition and presentation of unemployment statistics, almost all of which had the effect of reducing the so-called 'headline' figures. Some of these changes were justified in the interests of improving the precision and accuracy of the measurement of unemployment, but there are lingering doubts that the alterations were also intended to reduce the political impact of rising levels of unemployment in the early 1980s. At the time the Conservative Government was accused of 'fiddling the figures', particularly by the exclusion of certain groups from the category of unemployed. Whatever the motives behind these changes, they had the consequence of altering the base for the measurement of unemployment and made it more difficult to chart trends over time.

Other demographic issues which are currently the focus of political debate include changes in marriage and divorce and the role of the family in contemporary society. Figures which highlight the increasing levels of divorce, single-parent families and illegitimacy in the UK can be taken to imply a deterioration in moral standards and the demise of secure family life. Such figures need to be treated with caution as they portray only part of the picture of changing social behaviour (see Chapter 5). Clearly, statistical material can be selected and manipulated to support almost any argument, and published statistics should always be interpreted with some consideration of the purpose behind their collection and presentation.

This is not to suggest, of course, that all statistics produced by central government should be treated as propaganda. Clearly, the majority of the detailed information is uncontroversial and the agencies responsible for the collection, analysis and publication of data carry out their responsibilities conscientiously and meticulously. However, the objectives and traditions of the offices responsible for data collection have had a profound effect on the nature of research into social and demographic issues and on the interpretation of statistical material. In particular, the format and content of the census has had a major influence on the topics chosen for investigation and particularly on the derivation of demographic and social indicators. Since the census of 1841, information has been recorded about 'rank, status or occupation' and used to subdivide the population into different occupational classes. The Registrar General's definition of five principal socio-economic categories (1911 census) has become the established way of constructing social groupings and these groupings are frequently adopted in research without too much concern about the assumptions on which they are based (see Chapter 3). Similarly, notions of 'quality of life' and 'social deprivation' have been significantly influenced by the recording of information in the census on physical amenity (fixed baths, WCs, etc.), shared accommodation or access to cars. These are surrogate measures for housing conditions and relative wealth, and in many instances they provide only a limited or generalized view of the reality of the situation and do not effectively convey the differences in life-style and living conditions between different social groups.

When utilizing statistical information collected by central government to study contemporary social and demographic issues, it is important to remember that although the data may be systematically gathered, carefully processed and presented in a comprehensive manner, it may still provide only a partial view of the structures and trends within British society. There are many other facets of life which are inadequately recorded because they are not the subject of survey or because there is no provision for the registration of information. Hence there is a danger that the statistical requirements of government may determine the agenda for socio-demographic research and influence the conceptualization of social patterns.

# 3

# METHODS OF DEMOGRAPHIC ANALYSIS

Like all subjects, demography has a language and a methodology of its own that has developed in response to the need for ways of describing the characteristics of populations accurately and precisely. Over the years this has led to the establishment of a standard set of criteria for measuring such basic phenomena as fertility and mortality, and has provided a common frame of reference for the comparison of one population with another, or for assessing changes over time. The concepts of fertility and mortality are easy enough to grasp, but their precise definition and measurement require the use of analytical techniques which accommodate all the relevant factors. For example, measures of fertility need to take into consideration not just the number of children being born but also the size of the total population, or more specifically the size of the reproductive age-group. Measurements can be further refined to take account of the variations in child-bearing across the reproductive age-range and the impact of female age-specific mortality rates – all of which help to provide a more detailed statistical estimate of the *rate* at which children are being added to the population. Some of the techniques used require fairly complex procedures and at first sight may seem daunting to those not experienced in statistical analysis. But in almost all cases the methods are based on simple principles and it is not always necessary to understand the mechanics of calculation in order to appreciate the meaning of the results.

The great advantage of the development of standardized statistics for the measurement of populations is that they provide a comprehensive basis for analysis. The problem is that, in so doing, they necessarily determine the interpretation of the phenomena being measured. By applying different techniques for the measurement of fertility, for example, it is possible to present quite markedly different impressions of prevailing trends in a population. Similarly, demographers have established (or adopted) a number of ways of classifying groups within the population (socio-economic groups, ethnic groups, standard regions, etc.) in order to identify different dimensions to the explanation of population change. Many of these schemes for categorizing data are preserved because they provide a basis for measuring change over time. But in some respects they do not provide a very precise measure of the current situation. For example, there are several different ways of defining ethnic groups in the population, each of which may give a different impression about the size, dynamics and likely future trends of these groups. These points need to be borne in mind. The understanding of Britain's population is dependent not just on the availability of systematically collected data but also on the established ways in which this information is processed and presented.

## Measurements of population structure

There are two basic dimensions to the structure of any population, namely the divisions between the sexes and the divisions between different age-groups. It is conventional to represent these characteristics in the form of a population pyramid which identifies the relative distribution of numbers between age categories (Figure 3.1). Quite apart from providing a neat and easily interpreted picture of population structure, pyramids can also give information about the recent history of population dynamics and about the varying experiences of different cohorts within the population. For example, the peaks in fertility in Britain in the late 1940s and mid-1960s can be traced on the contemporary pyramid as bulges in the age-groups 40–44 and 25–29.

Changes over the longer term can be analysed through the construction of pyramids for successive time periods. The impact of variations in the prevailing trends of fertility and mortality results in different overall shapes for the pyramids. Populations that experience high levels of fertility and mortality (especially infant and child mortality) develop pyramids with a

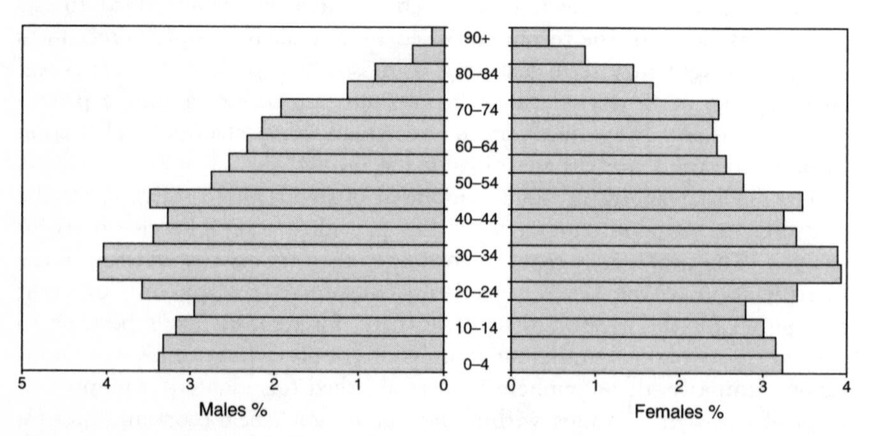


*Figure 3.1* Age/sex structure (England and Wales) 1994

*Source:* OPCS (1995), table 1

*Note:* Percentage of total population in each five-year age category wide base and a narrow apex. Mature populations, like Britain, with low levels of mortality, declining levels of fertility and a stable demographic regime, have pyramids which are more barrel-shaped; the majority of deaths are concentrated in the oldest age categories and the relative decline in births results in a narrowing of the base. It is also possible to identify likely trends in population change from the detailed shape of a pyramid, particularly the relative size of different age cohorts in the future. Figure 3.1 suggests that there will be a significant increase in the numbers reaching retirement age in Britain in the years 2021–5, for example.

Population pyramids are constructed with the age categories on a central vertical axis. Normally, one-year or five-year age categories are used and the final ages are collapsed into one category, e.g. 85 and above (for ageing populations this may include some misrepresentation of the actual distribution, as strictly speaking the figures should be spread over the remaining fifteen years or so of a possible life-span). Figures for males are recorded on the left of the diagram, with females on the right. The horizontal scale may either represent actual numbers (useful when representing changes in population size over time) or the percentage of the total population in each age category (useful for comparing differences in structure between populations of different sizes). The age categories may also be subdivided to show the component elements of a population as, for example, in Figure 3.2 which shows the population divided into four categories – single, married, divorced and widowed.

The division between the sexes can also be analysed by calculating the male:female (or sex) ratio. This expresses the number of men in the population as a proportion of the number of women (usually per 1,000 women). Minor differences in the sex ratio between different age-groups are difficult to detect on a population pyramid, but the calculation of age-specific sex ratios can identify significant changes in the relative proportions of men and women at different stages in the life-cycle (Figure 3.3). There is a consistent pattern to the sex ratio with relatively more boys being born than girls, but with women living longer lives. For every 1,000 female births in Britain there are approximately 1,050 male births. By the age of 35 the relatively higher male mortality rate results in a roughly even division between the sexes. Above the age of 55 the pattern swings dramatically in favour of women so that by the age of 85 the male:female ratio is in the region of 332 males per 1,000 females (*Population Trends* 85, table 6).

## Measurements of fertility

Measurements of fertility indicate the rate at which births are adding new members to a population. They are expressed in relation to the size of a population or to the number of women of child-bearing age. Patterns of fertility are influenced by many factors, some of which are purely functional,

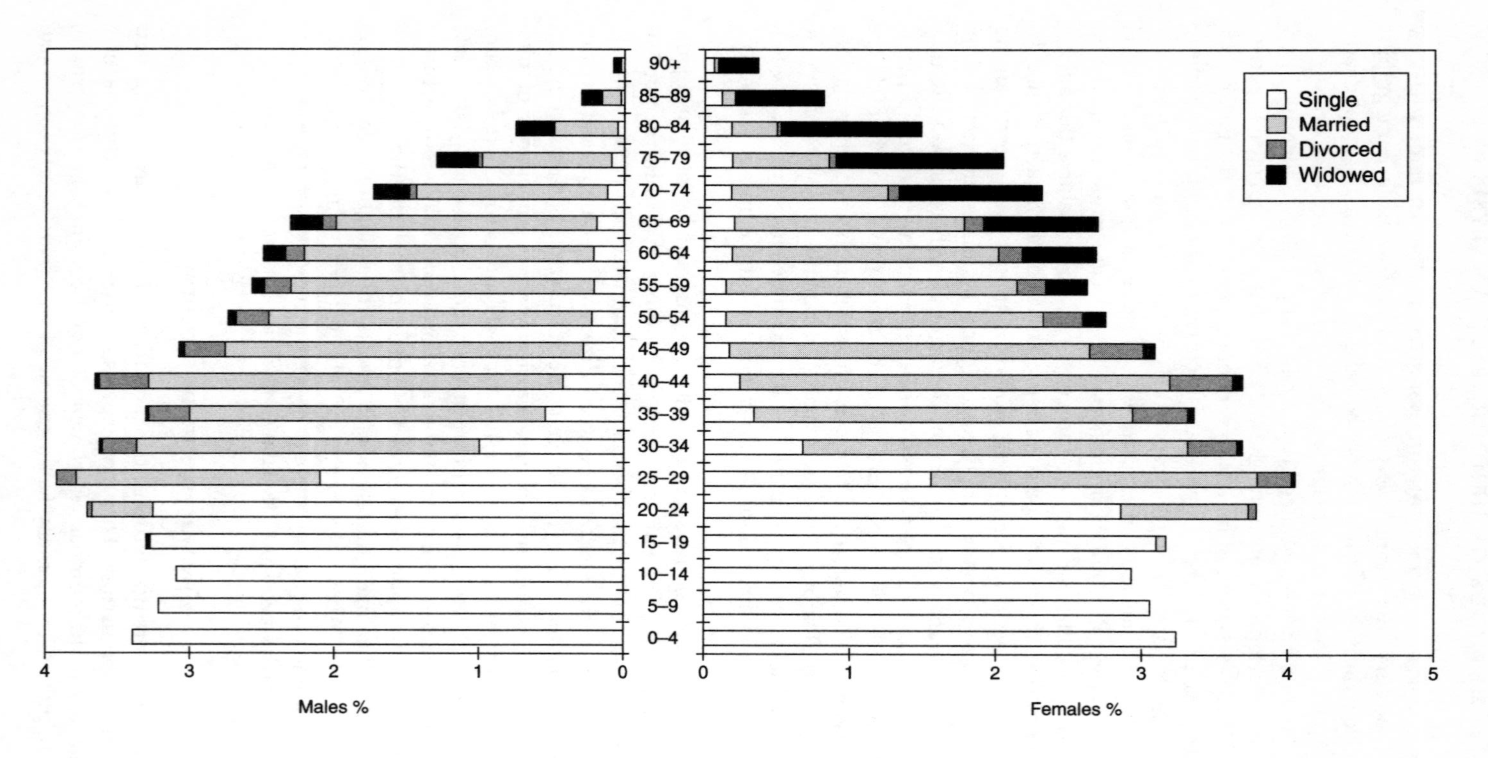


*Figure 3.2* Age/sex structure with marital status 1991

*Source:* OPCS (1993a), table 2

*Note:* Percentage of total population in each five-year age category

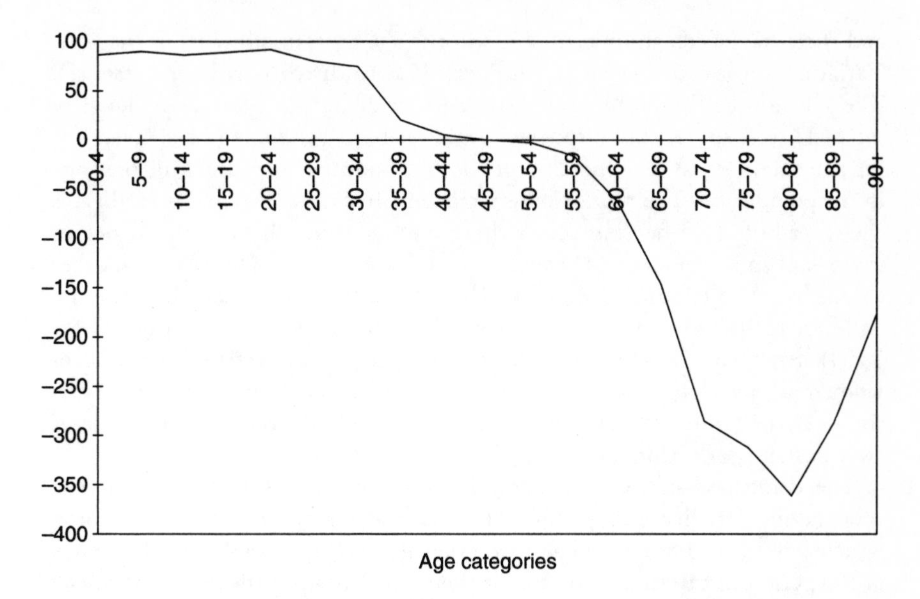


*Figure 3.3* Sex ratio (males per 1,000 females) by age category 1994

*Source:* OPCS (1995), table 1

such as the variations in age cohorts, but others are more subtle and difficult to identify, such as changing attitudes towards family size. Over the recent past, changes in the prevailing trends of fertility have been the most significant factor in determining overall population change in Britain and have created peaks and troughs in the age structure which will have reverberations for many years to come (see Chapter 4).

The simplest measure of fertility is the crude birth rate. This expresses the number of recorded births in a given year as a proportion of the estimated total mid-year population. The crude birth rate is a widely used statistic because of its general availability and ease of calculation. Even for countries with very rudimentary procedures for collecting population data, it is possible to arrive at estimates for the number of births and population size. However, it is imprecise and can be a very unreliable measure of actual fertility because it fails to take account of the age distribution of a population and particularly the size of the reproductive age-group in relation to the rest of the population.

A more precise measure of fertility can be obtained by dividing the total number of births by an estimate of the number of women in the reproductive age categories (normally taken as ages 15–49). However, this does not acknowledge the fact that child-bearing is unevenly spread across this age-range. The majority of children are born to women aged between 20 and 35 and there have been significant shifts in the age pattern of births in the past. Variations in the size of age cohorts can lead to distortions in the statistic. The solution to the problem is achieved by calculating age-specific fertility rates. These express the number of children born to women of a given age (or age category) as a proportion of the total number of women of that age in the population. The plot of age-specific fertility rates (Figure 3.4) indicates the spread of child-bearing across the age-range of mothers and provides an important indicator of one of the principal determinants of fertility. All other things being equal, during periods of high fertility there is tendency for more children to be born to women in the earlier reproductive years (figures for 1964). An early start to family formation extends the fertility potential of women. Conversely, one of the most powerful constraints on fertility is the delay of family formation and the shift of child-bearing into the later twenties and early thirties (figures for 1996).

The combined index of age-specific fertility rates is referred to as the total period fertility rate (TPFR). This is 'the average number of children which would be born if women experienced the age-specific fertility rates of the period in question throughout their child-bearing lifespan' (*Population Trends* 62, footnote to table 9). In other words, it is a more refined measure of the number of births per woman in the reproductive age categories, taking into account the differences in size of different age cohorts and the

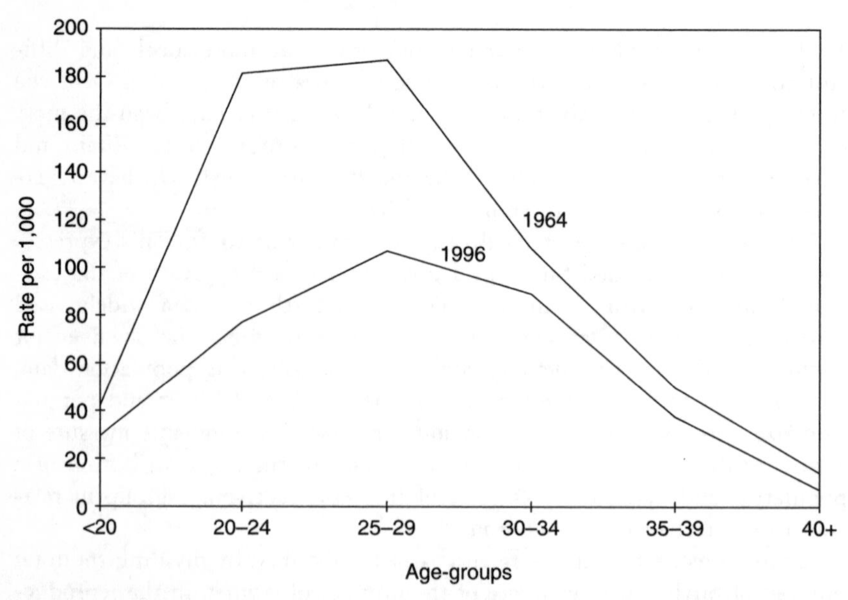


*Figure 3.4* Age-specific fertility 1964 and 1996. Births per 1,000 women in each age-group

*Source: Population Trends* 88, table 9

variations in age distribution of births. The TPFR is a commonly used measure and provides a sensitive index to prevailing trends in fertility (Figure 3.5), but its drawback is the assumption of stability in age-specific fertility over time. As noted there may be quite marked variations in the distribution of births by age cohort and it is unrealistic to assume, for example, that women of twenty today will in fifteen years' time experience the levels of fertility of the current generation of women in their mid-thirties.

A preferred method of estimating the actual fertility of women is the measurement of achieved family size (or completed fertility). This records the total number of children born during the reproductive life-span and therefore takes into account the variations over time in age-specific rates. It is essentially a measure of historic trends in fertility, indicating the experience of women over the previous thirty years. Hence it is useful for explaining changes in the past but is of less value for estimating the prevailing trends in fertility.

A more exact way of measuring fertility is to calculate the rate at which females replace themselves in the population. The net reproduction rate (NRR) is measured by recording the number of female children born and by taking account of the loss of women through death before the end of the reproductive period. Its calculation requires information on the number

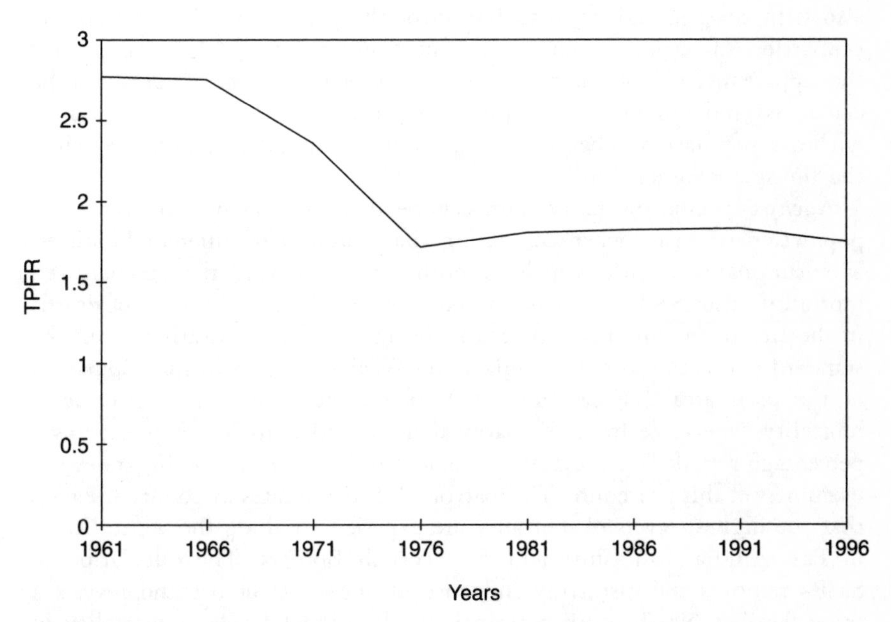


*Figure 3.5* Total period fertility rate 1961–96

*Source: Population Trends* 88, table 9

of female births to mothers in each age category and knowledge of the prevailing trends of female mortality (see life tables, page 35–6). The figures are expressed as a ratio of the number of women of reproductive age. Hence an NRR of 1.0 is indicative of a stable population. Figures in excess of one indicate the potential for growth if prevailing trends are maintained and those of less than one suggest a population is not achieving replacement (cannot sustain its present size in the future). The NRR for England and Wales in 1977 was 0.79 – the low point in postwar fertility (OPCS 1987a).

## Measurements of mortality

Measurements of mortality indicate the rate at which people are dying in a population. As with fertility, there are a number of statistical techniques which can be applied to identify the significance of deaths in relation to the structure of the population. In Britain the chances of dying are largely determined by age. In 1990 81 per cent of deaths in England and Wales were of people aged 65 and above (OPCS 1991b), the most common causes of death being heart disease, respiratory diseases and cancer.

The crude death rate, like the crude birth rate, is the simplest measure. The total number of deaths is expressed as a proportion of the mid-year estimates of population size. Also like the crude birth rate, this is an unreliable measure because it varies not only with the number of deaths but also with the age and sex distribution of the population. The greater the proportion of a population in the more elderly categories, the greater the apparent level of mortality. Age/sex-specific mortality rates can be calculated and provide a more precise indication of the level of mortality within a population. They present a profile of mortality experience across the life-span (Figure 3.6).

Age/sex-specific mortality rates can be used to iron out differences in population structure when comparing the mortality of different localities. The summary statistic which is normally applied is the standardized mortality ratio (SMR). This is 'the ratio of the observed number of deaths in the area to the number expected if the age/sex-specific death rates in the standard population (i.e. for England and Wales) applied to the population of the local area' (OPCS 1990: 13). It indicates the deviation of local mortality experience from the national norm and is usually expressed as a percentage rounded to the nearest whole number. Figure 3.7 illustrates the usefulness of this procedure. The map of crude death rates by county suggests that the highest levels of mortality are experienced along the South Coast, in East Anglia, Lancashire and the Scottish Borders. The calculation of SMRs removes the distorting influence of age/sex structure and reveals a more familiar North/South pattern with the higher levels of mortality in the older industrial regions of Northern and Western Britain and lower levels in the South and East.

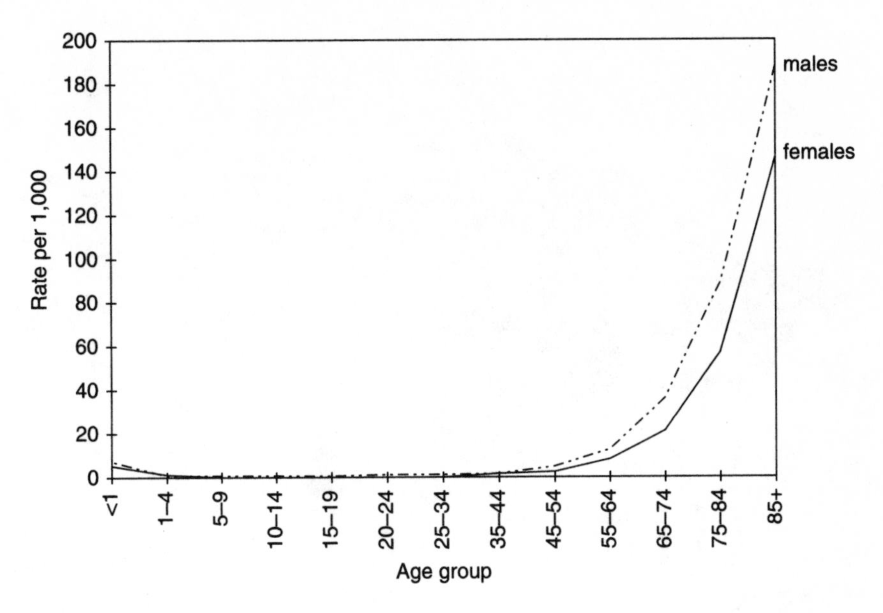


*Figure 3.6* Age/sex mortality rates 1994

*Source: Population Trends* 88, table 13

Certain categories of mortality are represented by specific statistics. The death of children in the early stages of life is a particularly sensitive indicator of general levels of public health and nutrition. The statistics are recorded for different time periods after birth, and are expressed in proportion to the number of live births. The infant mortality rate records the number of deaths of children aged under one year per 1,000 live births in a given year. The neonatal mortality rate records deaths in the first four weeks of life per 1,000 live births, and the perinatal mortality rate records deaths in the first week together with still-births per 1,000 total births (live births plus still-births). This effectively measures death at and around the time of birth. The significance of the distinction of these categories of mortality relates to the cause of death. The majority of deaths after the first few days of life are caused by exogenous factors (mainly disease), whereas neonatal and perinatal mortality is largely associated with endogenous causes (premature birth, congenital malformations, provision of obstetric and paediatric care, etc.). There have been dramatic improvements in the level of exogenous (or post-neonatal) mortality in the past, but neonatal mortality has been more resistant to reduction (Benjamin 1989).

An alternative to the measurement of the rate at which people die is to provide estimates of the chances of survival. Life-expectancy from birth is a commonly used statistic for summarizing general trends in mortality and

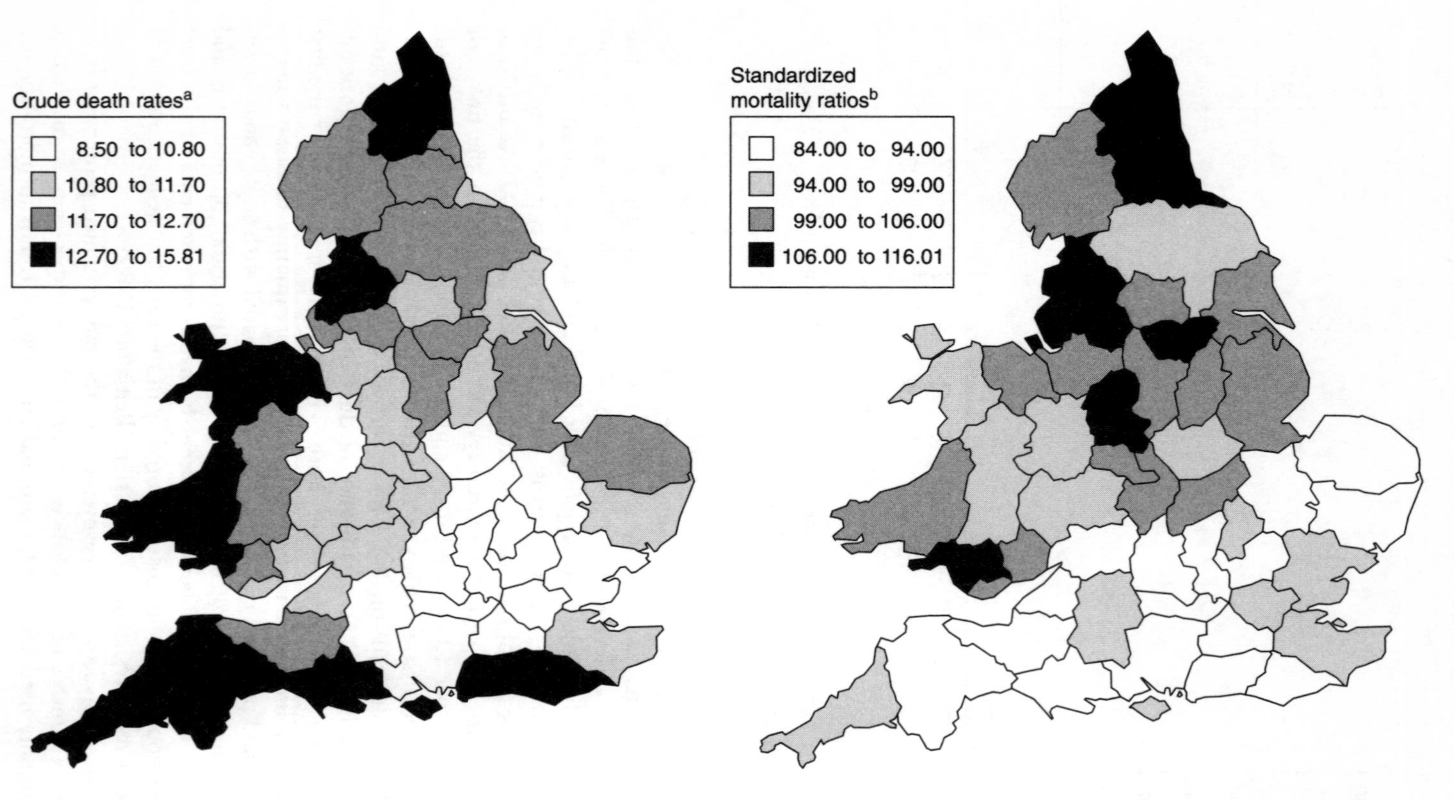


*Figure 3.7* Crude death rates and SMRs by county 1993

*Source: Key Population and Vital Statistics for Local and Health Authority Areas* (1994), table 4.2

*Notes:*

a deaths per 1,000 total population

b ratio of actual deaths to expected deaths for each county (see p. 32) varies with the changing age pattern of deaths. It is frequently interpreted as the average length of life, although this is not precisely what it represents. Life-expectancy varies during the life-cycle. At birth there is a probability of dying at every age. Once a period of life has been survived, then the chances of living longer increase. For example, life-expectancy at birth for males in 1994 was 73.9 years, whereas life-expectancy for those aged 30 was a further 45.2 years (*Population Trends* 86, table 12). It would also be inappropriate to estimate from these figures the likely age at death for an individual or cohort. They are based on prevailing levels of mortality (the current age-profile of death) and not on the levels of mortality which may pertain in the future when the upper-age categories are reached.

Estimates of life-expectancy are derived from life tables. These are summarized from mortality statistics of individuals at different ages. Because there are significant differences in the mortality of males and females, life tables are always presented separately for each sex. There is a certain mysticism about the construction and interpretation of life tables which derives largely from the use of algebraic notation and statistical formulae in their definition. But the principles on which they are based are quite straightforward. Given the knowledge of the age distribution of deaths, it is possible to derive precise estimates of the probability of an individual surviving for a given period of time in the future. For example, for a male aged 20 it is possible to calculate the likelihood of death before attaining the age of 40. The value of this information for the purposes of life assurance is clearly apparent.

Table 3.1 presents the abridged life table for England and Wales for 1986–88 prepared by the Registrar General. Information is provided for the population for the first five years of life and then at five-yearly intervals. The figures are expressed in proportional terms (per 10,000 individuals). Starting from birth, the table shows for each age point the numbers surviving between successive age points ($lx$) and the average remaining years of life ($ex$). It is a period table because it refers to the mortality patterns effective at one particular point in time (1986–1988). Given current age-specific mortality rates, of the 10,000 male births in 1986–8, 123 are likely to die before their fifth birthday and hence 9,877 will survive to exact age 5 ($lx$). Expectation of life from birth, for males, is 72.4 ($ex$). By reading down the $lx$ column it is possible to determine the ages by which proportions of a given cohort will die. In this case one-half of the male cohort will die before the age of 75 and four-fifths before the age of 85. Note that expectation of life is significantly longer for females.

Abridged life tables provide a summary view of mortality for any given year and are useful for identifying changes over time. For the demanding purposes of estimating precisely the size of the population entitled to state pensions at some date in the future, or calculating accurately the profit margins on life assurance premiums, it is necessary to work with more

*Table 3.1* Abridged life tables (England and Wales) 1986–88

| Age $x$[a] | Males $lx$[b] | $ex$[c] | Females $lx$[b] | $ex$[c] |
|---|---|---|---|---|
| 0 | 10,000 | 72.4 | 10,000 | 78.1 |
| 1 | 9,894 | 72.2 | 9,921 | 77.7 |
| 2 | 9,887 | 71.2 | 9,914 | 76.7 |
| 3 | 9,883 | 70.2 | 9,910 | 75.8 |
| 4 | 9,879 | 69.3 | 9,908 | 74.8 |
| 5 | 9,877 | 68.3 | 9,905 | 73.8 |
| 10 | 9,866 | 63.4 | 9,897 | 68.9 |
| 15 | 9,854 | 58.4 | 9,889 | 63.9 |
| 20 | 9,820 | 53.6 | 9,875 | 59.0 |
| 25 | 9,779 | 48.8 | 9,859 | 54.1 |
| 30 | 9,740 | 44.0 | 9,841 | 49.2 |
| 35 | 9,692 | 39.2 | 9,813 | 44.3 |
| 40 | 9,630 | 34.5 | 9,770 | 39.5 |
| 45 | 9,530 | 29.8 | 9,701 | 34.8 |
| 50 | 9,354 | 25.3 | 9,584 | 30.2 |
| 55 | 9,055 | 21.0 | 9,396 | 25.7 |
| 60 | 8,540 | 17.2 | 9,086 | 21.5 |
| 65 | 7,706 | 13.7 | 8,581 | 17.6 |
| 70 | 6,527 | 10.7 | 7,854 | 14.0 |
| 75 | 4,997 | 8.2 | 6,786 | 10.8 |
| 80 | 3,277 | 6.2 | 5,347 | 8.0 |
| 85 | 1,720 | 4.6 | 3,547 | 5.7 |

*Source*: *Mortality Statistics,* OPCS Series DH1, no. 21 (1990), table 15

*Notes:*
*a* $x$ = age point
*b* $lx$ = nos surviving to successive age points
*c* $ex$ = average remaining years of life

detailed life tables which specify the parameters for each individual year of life across the age-span. Age survival rates are expressed in terms of exact years (the number who survive for 365 days – from exact age $x$ to $x + 1$) and correction factors are built in to account for other factors such as the seasonal variations in deaths and short-term fluctuations caused by harsh winters or flu epidemics. Full life tables are produced every ten years by the Government Actuary, based on census figures and the mortality statistics in the three-year period spanning the census. Insurance actuaries prepare life tables on a more regular basis and over time have built up a very detailed picture of the mortality statistics of different groups within the British population (e.g. specific occupational groups, smokers and non-smokers, etc.).

## Migration statistics

Measurements of fertility and mortality are based on systematically recorded vital events and calculated with a high degree of precision. The measurement of migration is less clear-cut because the term only refers to a sub-set of population movements which are not monitored continuously and which may be subject to variations in interpretation. Generally, migration is taken to represent a permanent change of residence which involves the crossing of an administrative boundary. Migrants are identified in the census as those living in a different local authority area one year prior to the date of the census. The NHSCR defines migrants as those who register with a different Family Health Services authority. Both definitions provide imprecise estimates of the true level of migration, but are generally accepted to indicate the prevailing trends in population movement (see Chapter 2).

The simplest way of measuring the significance of migration for any given area is to calculate the net migration. This is the contribution to population change not accounted for by births and deaths. If precise figures for population size are known for two dates (for example, census years), and there is an accurate record of the number of births and deaths for the intervening period, then the difference between natural change (births/deaths) and actual change must be accounted for by migration. This provides an indication of the contribution of migration to population change. It does not provide a very clear estimate of the actual level of population movement because net values only indicate the difference between inflow and outflow and not the scale of movement in each direction.

Migration estimates based on recorded movements give a more detailed picture. Table 3.2 is taken from the ONS publication *Key Population and Vital Statistics for Local and Health Authority Areas* and is based on National Health Service data. For each area it is possible to provide estimates for movements in and out, and net migration (expressed as net inflow) for males and females and for different age-groups. The migration rate for each area is the net migration expressed in relation to population size (per 1,000 resident population). This detailed information is summarized at the regional scale as a matrix of movement between areas of origin and areas of destination (Table 3.3). The figures on the diagonal represent moves between Family Health Services authorities within each region. The row totals give the gains from migration from all other regions, and the column totals give the losses. The difference between the two is the net migration for each region in relation to all the other regions in England and Wales. There are, of course, other elements to net regional migration represented by flows to and from other parts of the UK and from immigration and emigration. Migration matrices such as this give an indication of the basic patterns of inter-regional population movement. It is possible to extend the techniques of matrix analysis to measure other aspects of migration and to predict likely future trends (Woods 1979).

*Table 3.2* Net migration by age and sex 1994 South East

| Area | Persons (thousands) | Males | Females | 0–14 | 15–29 | 30–59 | 60+ |
|---|---|---|---|---|---|---|---|
| South East | | | | | | | |
| In | 216.6 | 104.4 | 112.2 | 29.1 | 121.5 | 52.5 | 13.5 |
| Out | 231.4 | 112.3 | 119.1 | 35.6 | 105.5 | 68.3 | 22.0 |
| Net | –14.7 | –7.9 | –6.9 | –6.5 | +16.1 | –15.8 | –8.5 |
| Greater London | | | | | | | |
| In | 151.8 | 73.0 | 78.9 | 15.2 | 94.0 | 36.5 | 6.1 |
| Out | 197.2 | 93.4 | 103.9 | 31.7 | 76.8 | 66.8 | 22.0 |
| Net | –45.4 | –20.4 | –25.0 | –16.5 | +17.2 | –30.3 | –15.8 |
| Remainder of South East | | | | | | | |
| In | 259.4 | 123.8 | 135.6 | 42.9 | 112.2 | 79.0 | 25.4 |
| Out | 228.8 | 111.3 | 117.5 | 32.9 | 113.3 | 64.5 | 18.1 |
| Net | +30.6 | +12.5 | +18.1 | +10.0 | –1.1 | +14.5 | +7.3 |

*Source: Key Population and Vital Statistics for Local and Health Authority Areas* (1994), table 5.1

*Notes:* The South East is the standard planning region; Greater London and the Remainder of the South East are subdivisions to differentiate between the capital and outer region

*Table 3.3* Migration between standard regions 1994

| Area of destination | Area of origin (thousands) | | | | | | | | |
|---|---|---|---|---|---|---|---|---|---|
| | North | Yorks & Hum. | East Mids | East Anglia | South East | South West | West Mids | North West | Wales |
| North | 33.8 | 9.3 | 3.6 | 1.6 | 11.4 | 2.6 | 3.2 | 7.9 | 1.3 |
| Yorks & Hum. | 10.7 | 57.7 | 13.8 | 3.6 | 22.5 | 5.6 | 7.5 | 15.7 | 2.8 |
| East Mids | 4.2 | 16.6 | 28.2 | 7.3 | 30.8 | 6.6 | 13.7 | 9.8 | 3.2 |
| East Anglia | 1.9 | 4.2 | 7.1 | 12.4 | 32.4 | 4.0 | 3.4 | 3.4 | 1.5 |
| South East | 13.2 | 24.0 | 27.3 | 24.3 | 566.2 | 53.5 | 29.3 | 29.6 | 15.5 |
| South West | 3.2 | 7.2 | 7.8 | 4.4 | 67.1 | 50.3 | 14.4 | 9.4 | 8.8 |
| West Mids | 3.2 | 7.2 | 11.2 | 2.8 | 26.2 | 11.6 | 73.7 | 11.5 | 7.5 |
| North West | 8.3 | 15.1 | 8.4 | 2.9 | 25.6 | 7.1 | 11.2 | 95.0 | 7.8 |
| Wales | 1.4 | 2.8 | 3.1 | 1.4 | 15.3 | 8.3 | 8.2 | 9.6 | 24.5 |

*Source: Key Population and Vital Statistics for Local and Health Authority Areas* (1994), table 5.2b

## Population estimates and projections

The collection of comprehensive information about the size and structure of the total population only occurs every ten years at the time of the census. For other years it is necessary to calculate estimates of population from the available information about population change. Detailed mid-year estimates of population, by local authority and health district, are required by the Department of the Environment and the Department of Health for resource allocation, and are widely used by other government departments for monitoring changes in the population. In addition, there are many applications of the information, by local and central government, for planning and administration. The estimates also form a necessary component for calculating many of the basic statistical indicators of population change (e.g. birth rates and death rates).

The ONS produces annual estimates of the resident mid-year population (as of 30th June) for England and Wales. These estimates are based on the figures from the previous census with allowance for births, deaths, migration and ageing of the population over the intervening period. Summaries of the components of change at the national level (England and Wales) are provided for each year since the previous census (figures are also given separately for England and Wales). These identify the relative contributions of fertility, mortality and migration to overall trends in population change. Estimates are also provided for the age and sex structure of the populations for five-year age categories and for each single year of age (up to age 89). The mid-year estimates for 1994 have been used to produce the age/sex pyramid in Figure 3.1.

At a more detailed level, information is published for London boroughs, counties and districts, and health areas. The tables identify the mid-year population for the base year (i.e. the preceding census year), the previous year and the current year (Table 3.4). Change since the base year is given as a total figure and as a percentage, and the components of change are identified separately (natural change and migration). Obtaining precise estimates of actual population change at this level is a complex task and clearly the greater the period of time that has elapsed since the last census, the less accurate the estimates are likely to be. However, over the years the ONS has developed a reliable methodology for obtaining population estimates. When estimates were rebased in 1981 (i.e. recalculated on the basis of the new census returns), the average error for total population of local authority areas was only approximately 2.5 per cent. Such a figure is reassuring, although for certain areas (e.g. inner cities) the degree of error was significantly higher.

For census years the enumerated population count is recalculated to a mid-year estimate to take account of the time-lapse between the date of enumeration and 30th June. There are other slight differences between the

*Table 3.4* Mid-year population estimates 1994

| *Area* | *Mid-year population (thousands)* | | | *Change (%)* |
|---|---|---|---|---|
| | *1991* | *1993* | *1994* | *1991–4* |
| England and Wales | 51,099.6 | 51,439.2 | 51,620.5 | 1.0 |
| England | 48,208.1 | 48,532.7 | 48,707.5 | 1.0 |
| Wales | 2,891.5 | 2,906.5 | 2,913.0 | 0.7 |
| Standard regions | | | | |
| North | 3,091.7 | 3,102.3 | 3,099.9 | 0.3 |
| Yorks and Humberside | 4,982.8 | 5,014.1 | 5,025.8 | 0.9 |
| East Midlands | 4,035.4 | 4,082.9 | 4,102.5 | 1.7 |
| East Anglia | 2,081.9 | 2,093.9 | 2,105.6 | 1.1 |
| South East | 17,636.8 | 17,769.4 | 17,870.6 | 1.3 |
| South West | 4,717.8 | 4,768.0 | 4,795.7 | 1.7 |
| West Midlands | 5,265.5 | 5,289.7 | 5,295.0 | 0.6 |
| North West | 6,396.1 | 6,412.4 | 6,412.1 | 0.2 |

*Source:* OPCS (1995)

base population used for estimates and the census population. Students are included as residents at their term-time address and members of HM and non-UK armed forces in England and Wales are included at their station. HM forces stationed outside the country are not included. Children at boarding school or in care, inmates of prisons (with sentences over six months) and long-stay patients in hospital, are similarly allocated to the area of the institution rather than their home address. The details of births and deaths for local authority and health authority areas are available from civil registration returns. Allowance is made for the gap between the events and their recording (for example, by tracking births from the end of one year in the registrations of the first month of the following year). Information on migration between local areas is obtained from the NHSCR (see above and Chapter 2).

The figures for estimated populations are generally published in the year after the date to which they relate. In order to provide more up-to-date information for local and central government planning, the ONS has produced sets of extrapolated estimates for each local and health authority in England. These effectively roll-forward the current estimates for the following two years, assuming that each local area maintains its share of national births, deaths and migration. Although this may be problematic in some localities where demographic patterns are less stable (e.g. inner-city areas), a comparison of extrapolated estimates with the canvass for the ill-fated Community Charge Register in 1989 suggested that the margins of error were very small (Rowntree 1990).

Estimated mid-year populations are used as the base for population projections. National projections are prepared by the Government Actuary in association with the Registrar Generals for England and Wales, Scotland and Northern Ireland. They provide an estimate of the future population of the UK, based on a set of assumptions derived from current demographic statistics. The primary purpose of projections is to provide a reliable and comprehensive data set for longer-term planning, particularly for the provision of pensions and other forms of retirement income, although they are widely used by government departments and local authorities as an authoritative base for forward planning. The value of these projections is not so much their accuracy – past experience suggests that predicting the future can be a hazardous business – but the fact that they are based on consistent assumptions. They provide 'a view of the future that, in an uncertain world, seemed to be most likely at the time of preparation, rather than a confident forecast of what will actually occur' (OPCS 1991d: 1).

National projections are made every two years with interim annual revisions. The principal period of projection is 40 years, although figures for the UK and Britain are rolled forward for a further 30 years to provide estimates of population size and structure for longer-term needs (e.g. the management of pension funds). Basic tables, showing projections at five-year intervals for males and females in each age category, are published for the UK, Britain, England and Wales, Scotland and Northern Ireland. More detailed information is included on microfiche; this contains tables summarizing the annual components of change and gives projected populations by individual ages, as well as the variant projections, using alternative assumptions about fertility and mortality rates.

The methodology of population projection is similar to that of population estimates. It is relatively simple in principle, but complex in terms of the statistical calculations involved. The technique is referred to as the component method and involves information about four principal factors: the base population, levels of mortality, levels of fertility, and migration. The detailed information is organized on computer spreadsheets and the analysis involves a year-by-year estimation of changes for each single year of age. Starting from the base population on 1st July, losses from death and out-migration are subtracted from the figures for each age and gains through birth and in-migration are added. This gives a revised base population for 30th June in the following year.

Like all forecasts the accuracy of population projections is influenced by the assumptions that are made about future trends, in this case the likely variations in population dynamics. Short-term projections can be based on the prevailing levels of mortality, fertility and migration, but with increasing distance form the present such parameters become less and less reliable. Assumptions need to be made about likely future variations in these parameters and also about the rate and timing of changes. In the case of mortality, the general assumption is that rates will decline for most ages over the period of the projection. The rate of decline is calculated as a geometric progression, halving every ten years, so that after forty years the rate of improvement is relatively marginal. In recent years this stable pattern of mortality improvement has been interrupted by the impact of AIDS, particularly for men in the age-group 30 to 50. It is impossible to estimate the long-term effects of this disease, but informed assumptions have been built into the projections to make some allowance for it. This is a good illustration of the sort of problem encountered with projections. They can only be based on what is currently known and cannot take account of unforeseen events.

Estimates of births are based on completed generation fertility for women at each age between 15 and 46. Generation fertility is based on the year of birth of women and measures the total number of children a woman will bear during her reproductive life-time in comparison to the size of her generation (i.e. all those surviving females born in the same year). It is a more suitable measure than the total period fertility rate as it is not influenced by short-term variations in the age of women at the commencement and completion of family formation. Information about completed family size is derived from the *General Household Survey*. At present this is approximately two children per woman, about 5 per cent below the level required to replace the population from one generation to the next (OPCS 1989). Little change is expected in this pattern for the foreseeable future. However, levels of fertility tend to vary more significantly than mortality over the shorter term. Fertility is responsive to the influence of non-demographic factors, such as economic trends. It is conceivable that the assumptions of little change in prevailing levels of fertility may be overturned by events, as indeed they were in the early 1960s (see Chapter 4).

Information about patterns of migration into and out of the UK is taken from the *International Passenger Survey* (see Chapter 2). This provides details of the sex and age structure of migrants. Figures for movement between the four countries of the UK are derived from the NHSCR. In recent years international migration has only had a very limited impact on population change. Inflow and outflow have roughly been in balance with each other, with a net outflow during the 1970s matched by inflows in the period since 1983. For projections from 1989–90 to 2026–7 it is assumed that migration will have no impact on overall population change (i.e. net inflow = 0). However, differences in the age and sex characteristics of immigrants and emigrants have been taken into account. Generally speaking, immigrants are younger than emigrants and the projections assume a net inflow of young adults under the age of 25 and a counterbalancing net outflow of adults aged 25–34. Net migrational flows between England and Wales, Scotland and Northern Ireland have also been relatively consistent during the 1980s. England and Wales have made marginal gains at the expense of losses from Scotland and Northern Ireland. It is assumed that this pattern will remain unaltered in the future.

Every attempt is made to ensure that population projections provide the most realistic estimate possible of future trends. But it would be wrong to regard the figures as accurate forecasts of events. The ONS accepts that the principal projection presents only one view of what may happen. In order to take account of plausible changes in current trends, alternative – or variant – projections are also produced. Different assumptions about fertility, mortality and migration are used to produce high and low variants. These 'what if' scenarios present the opportunity of considering alternative futures and work through the implications of changes in terms of the resulting size and structure of the population.

In the case of fertility, assumptions for the lower-variant projections suggest a continuing fall in average completed-family size to a figure of 1.75 children for the generation of women born in 1965, followed by a marginal rise to 1.81 fifteen years later. These figures are similar to those achieved in Britain at the low point of fertility in the depression years of the 1930s. The higher variant projection assumes a gradual increase in fertility to approximately 2.2 children per family, significantly higher than at present but still lower than the levels of fertility achieved during the 1960s.

The variations in fertility are more marked, and have a far greater long-term impact, than changes in mortality. As noted earlier, Britain has a stable mortality pattern and further reductions in death rates are likely to be achieved only fairly gradually. Despite concern over the spread of AIDS, it is not expected that this disease will add significantly to overall mortality levels in the next century and, unless other unknown fatal diseases make an appearance in the future, it is highly unlikely that the long-term downward trends in mortality will be reversed. For this reason the variant projections for mortality assume only differing rates of mortality decline and not the possibility of a rate of mortality higher than at present. The lower variant is based on a rate of decline assumed to be only half as much as the principal projection. The higher variant uses a rate of decline 1.5 times as great. No other variables are built into the projections. The main effect of these variant assumptions is seen in the numbers of people over the age of 60 in the population. The high variant results in an additional 949,000 elderly people by the year 2027, but only an additional 163,000 under the age of 60.

Some indication of the relative contribution of fluctuations in fertility and mortality to overall population change can be seen in the differences between the high and low variants for each factor. The difference in projected population numbers, in 2027, between the high and low variant for fertility, is approximately 8 million people. That is, the high variant would add 3,582,000 more than the principal projection and the low variant would produce 4,388,000 less. The difference between the high and low variants for mortality is only around 2 million (OPCS 1989).

Changing the assumptions about the net inflow of migrants to the UK not only adds or subtracts numbers to the total population but also has a significant impact on the size of younger age categories and hence on potential fertility. The variant projection for migration is achieved by taking the pattern in 1985 when there was a net outflow from migration. By maintaining this assumption over a forty-year period, the total impact on numbers is quite marked particularly in the younger age categories. The variant assumption results in a net annual loss from migration of 17,300. When compounded, and the consequent loss of births is taken into account, the result is a total loss of nearly 500,000 (1 per cent of the UK population) by the year 2021. There is no reason to believe that this estimate is any better than that achieved from the assumption of a zero figure for net inflow. In practice the pattern of inflow and outflow is likely to vary in the future, much as it has done in the recent past.

In addition to the national projections, the ONS also publishes projections for local authority and health authority areas in England (*Population and Health Monitor Series PP3*). Sub-national projections for Wales and Scotland are published by the Welsh Office and the General Register Office for Scotland. These projections are based on the mid-year estimates of population and are provided for a more limited time period than national projections. Figures produced in 1991 were based on 1989 estimates and were projected forward to 1996, 2001 and 2011. For local authority areas, information is provided for English regions, counties, metropolitan districts and London boroughs. A separate set of projections is provided for regional and district health authorities (although substantially derived from the same data). Assumptions about general trends in fertility, mortality and international migration are taken from national projections. Figures for local areas are derived by taking the average deviations between local and national rates, for the preceding three years. These differences are assumed to remain constant in the future. Assumptions about local patterns of migration are provided by the Department of the Environment (DoE). They are based on the patterns of internal movement identified in the previous census, together with the figures used for population estimates. Some modifications are made following consultations with local planning authorities and the National Health Service. The DoE has developed a sophisticated model for local migration which provides estimates of both gross and net migration rates for each local authority area (Armitage 1986). However, changing patterns of migration are difficult to predict with precision at this level and are subject to a significant degree of variation over time. For this reason the projections for the London boroughs and the metropolitan districts, which experience relatively high levels of local movement, are considered to be less reliable than the figures for the shire counties (OPCS 1991d).

## Population classifications

The methods of demographic analysis discussed so far relate specifically to statistical information collected for births, deaths and migration. They illustrate ways in which current and future trends in population change can be identified and measured with accuracy and precision. In order to achieve this, the data may be grouped or classified according to simple criteria. The population may be subdivided into males and females, or into five-year age categories, for example. However, to progress beyond purely demographic analysis and search for explanations for trends in population change, it is necessary to examine other characteristics of the population. Issues such as where people live, their occupations, wealth, life-style and social status all have some bearing on the variations in demographic behaviour between different individuals and groups in society.

Social class variations perhaps provide the most useful example. It is generally well known that there are observable differences in demographic characteristics between the extremes of the social spectrum. Measures such as infant mortality, completed family size and life-expectancy, when cross-tabulated with social class, highlight differences in life-style and life-chances and identify different types of demographic experience within British society. In broad terms it is possible to speculate about the typical characteristics of these groupings. Middle-class couples, for example, tend to display different patterns of family formation to low income families, with the delay of child-birth until later in marriage and generally smaller completed families.

Much of this supposed variation may be due as much to general impressions as to detailed statistical analysis. Part of the problem relates to the definition of groupings and the establishment of a common frame of reference for measuring social variations. Social class is a concept which may be interpreted differently by different individuals. It not only applies to objective factors such as occupation and status but also to more subjective factors such as perceptions of identity and community and class consciousness. For the purposes of examining demographic patterns, the classification of social class is generally taken as a surrogate measure for observable differences in the life-style of different groups in the population. Where life-style may be taken as the sum total of the facets and characteristics of everyday life, different patterns of life create different types of demographic experience.

For demographic statistics, the most commonly used classification of social types is the Registrar General's socio-economic groups. This classification was first introduced in the reports for the 1911 census as a means for identifying the social class differences in mortality. Since then the classification has been refined (most notably in 1951) and the basic categories have been subdivided into more specific groupings to give a total of seventeen social types with additional groups (such as students) not included

*Table 3.5* Registrar General's socio-economic groups

| | |
|---|---|
| Non-manual | |
| I | Professional occupations |
| II | Intermediate occupations (including most managerial and senior administrative occupations) |
| IIIn | Skilled occupations (non-manual) |
| Manual | |
| IIIm | Skilled occupations (manual) |
| IV | Partly skilled occupations |
| V | Unskilled occupations |
| Other | |
| | Residual groups including, for example, armed forces, students, and those whose occupation was inadequately described |

*Source:* OPCS (1987a)

within the ranking (Table 3.5). Despite these refinements, much of the basic cross-tabulation of demographic data is achieved using the broad categories of social class.

The definition of socio-economic groups and social classes is derived essentially from one variable – the recorded occupations of males in the census or other statistical sources. Descriptions of occupations are obtained from the Registrar General's *Classification of Occupations* which is published periodically. This is based on the *Standard Industrial Classification* which defines, in some detail, each sector of activity in the British economy. Occupations are allocated to social classes on the basis of the type of work done or the level of skill involved, the dividing line between 'middle class' and 'working class' falling in the centre of the very large class III category.

The advantages of this classification of British society are simply the fact that it has been used extensively in the past and that it provides a standardized way of defining groups within society, which is applied by government departments, local and health authorities, academic researchers and others. For many reasons it cannot be accepted as a representative concept of the groups that actually exist in society. It is based on the assumption that the occupations of men determine the social status of households and some of the divisions that have been established are based on fairly arbitrary criteria (e.g. the difference between professional and intermediate occupations, or skilled and partly skilled manual occupations). Still less can it be said to provide an indication of the different styles of life which are associated with different groups in society. Other possible social indicators, such as wealth, home ownership, education or health, would produce groupings that would not sit conformably on top of those defined by occupation. In addition, the groups produced are very large and contain within them a high degree of internal variation. This is particularly true of class III which contains a large proportion of the total population (above 50 per cent).

There are many other specific problems with using this classification of social class such as the exclusion of certain social groups (students, the retired, armed forces, etc.) who end up as status-less citizens. All these factors need to be borne in mind when interpreting any statistical information presented by social class. The results obtained are in part dependent on the definitions used in the classification and it would be wrong to read too much into the observed differences in terms of assumed patterns of life. For example, Table 3.6 shows variations in completed family size for classes I–V. The observed differences between class IIIn and class IIIm could be a result of different patterns of family formation between middle-class and working-class families, but it could equally be a product of the way in which this class has been subdivided in the first place.

More recently, alternative definitions of social groups have been developed for identifying different types of residential area. This information is valuable for commercial activities such as market analysis and targeted direct mail. The whole procedure of identifying the profile of groups of people living in different areas is referred to as 'geodemographics' and there has been a significant expansion in this field in recent years. A number of companies now produce classifications of small-area census data which can be used to identify the location of different social groups (CACI ACORN system, Pinpoint PiN system, CDMS superprofiles). The methodology involves using variables derived from the census and other sources (electoral registers, financial data, etc.) to determine the characteristics of each census enumeration district or postcode area. The detailed descriptions for small areas are usually aggregated into a number of distinctive neighbourhood types. For example, the ACORN classification identifies thirty-eight different neighbourhood types which combine the geographical characteristics of areas with the socio-demographic characteristics of the people who live there. In addition, a separate ACORN classification has been developed for 'life-style groups' (Table 3.7). This gives a breakdown of the population

*Table 3.6* Social class variations in family size 1981

| *Social class* | *Family size*[a] *(%)* | | | |
|---|---|---|---|---|
| | 0 | 1 | 2 | 3+ |
| I | 44.9 | 18.3 | 26.7 | 10.1 |
| II | 49.8 | 18.5 | 23.6 | 8.1 |
| IIIn | 56.4 | 17.6 | 19.7 | 6.3 |
| IIIm | 48.5 | 19.6 | 22.2 | 9.7 |
| IV | 56.6 | 17.4 | 17.2 | 8.9 |
| V | 58.0 | 16.0 | 15.5 | 10.5 |

*Source:* Kelsall (1989: 119)

*Note: a* number of children per family

*Table 3.7* ACORN life-style groups

| | |
|---|---|
| LA | Rural singles |
| LB | Younger rural couples and families |
| LC | Older rural couples and families |
| LG | Younger traditional suburban couples |
| LH | Older traditional suburban couples and families |
| LI | Very affluent suburban younger couples and families |
| LM | Younger couples in council areas |
| LN | Older couples in council areas |
| LO | Adult families in council areas |
| LS | Younger urban singles |
| LT | Older urban singles |
| LU | Young urban couples and families |

*Source:* Flowerdew (1991: 37)

into social groups based on three principal factors: location (rural, urban, suburban), wealth (income, home ownership, consumer durables), and life-cycle stage (young married couples, older singles, etc.).

As with other social classifications, this methodology has its problems. It is primarily concerned with the description of residential areas and not with the ranking of 'classes' in society. Also, because of the confidentiality of information about individuals in the census, the categories refer to groups of people and not individuals. As a result, all those living in an area that has been ascribed a particular designation are assumed to have the characteristics of the majority group, although clearly not everybody living in a 'younger urban singles' locality would fit the image or life-style implied by the label. Increasingly, data firms are putting together individual-level data bases derived from mail-order purchases, electoral registers and subscription lists. These are likely to supersede residential area classifications because of their enhanced precision (Flowerdew 1991). Despite the limitations of residential area profiles, they still represent a significant step forward from the Registrar General's classification. They attempt to identify significant variations in the life-style of households and recognize the importance of variables such as location and life-cycle stage. In addition, because they are derived from the census, they offer a number of possibilities for the cross-referencing of demographic and social indicators in order to provide a more sensitive insight into the significance of variations in life-style on demographic factors (for example, studies of differences in mortality or family formation between life-style groups etc.). The ACORN classification, referred to above, has been included in the ONS *Longitudinal Study*. Area-based descriptors have been attached to individual records (see Chapter 2). These descriptors provide information about the area in which people live rather than ascribing the characteristics of a locality to the individuals in the study.

Other examples of the way in which methods of classification may influence the interpretation of demographic patterns include the study of ethic minority populations. Before the inclusion of a question on ethnicity in the 1991 census (see Chapter 2), the details of the size and structure of ethnic groups were a matter of some speculation. First and second generation migrants could be picked out from census questions about birthplace, but later generations were statistically indistinct from the rest of the population. Clearly, classifications based on birthplace alone provided an under-estimation of these groupings and failed to identify any process of social or spatial assimilation. The *Labour Force Survey* asks respondents to identify the ethnic group to which they belong and over a number of years has compiled sufficient statistical information to analyse some of the social and demographic characteristics of particular groups. Unfortunately the categories used are not identical to those introduced in the 1991 census. Hence these two sources of information are identifying different groupings. The inconsistency in the recording of minority populations raises questions about the distinctiveness and identity of groups. Respondents to the census and the *Labour Force Survey* are asked which category, from a given list, they fit in to. This is not quite the same as asking for their own description of ethnicity or how strongly they feel associated to any particular group.

It is not only the understanding of structures within society that may be influenced by the methods used for classifying individuals. Spatial classifications based on the use of administrative areas may also have a significant bearing on the interpretation of socio-demographic characteristics of particular places. Most of the geographically referenced information published by the ONS refers to the countries of the UK, standard planning regions, counties, local authority districts or health authority areas. These spatial frameworks may not always be the most appropriate way of carving up bits of the country in order to investigate the significance of location on demographic behaviour. The images of a North/South divide in Britain during the 1980s, for example, were derived in part from data presented by standard regions in publications such as *Regional Trends*. These areas have little functional significance and are simply agglomerations of counties, but the representation of data for these large areas has had the effect of masking other important spatial dimensions to social and economic change such as the distinction between the experience of towns and rural areas. One exception to the general representation of data by administrative area is the publication of data, by the Department for Education and Employment, by 'travel to work areas'. Although there is some debate about the significance of the boundaries around these areas, it does at least represent an attempt to relate the statistical data to an appropriate geographical area. Travel to work areas may be assumed to approximate to the spatial extent of local labour markets.

This chapter has introduced methods of demographic analysis and discussed some of the issues concerning the description and presentation of population data. The aim has partly been to demonstrate the use of these techniques and partly to raise questions about the ways in which the established methodology influences the understanding of demographic issues. Methods of analysis are not simply concerned with mathematical calculations and the application of complex statistical formulae. They define the way in which populations are measured and evaluated. Just as in the previous chapter it was suggested that the format of the primary data sources (particularly the census) has had a significant influence over the choice of topics for research in demography, so also it may be argued that the development of the language and methodology of demography has perpetuated a particular approach to the subject and determined the understanding of the contemporary condition of Britain's population.

The discussion of methods of classification has highlighted the point that data are not objective and value free, but subject to the application of preconceptions and assumptions about, for example, the inherent socio-economic characteristics of the population. There are many other areas where decisions and assumptions condition the analysis of data. The measurement of migration is influenced by established assumptions about the duration of stay and the types of movement which constitute migration. Different sets of assumptions might present a different type of picture. For example, if the definition of migrants as 'those who cross an administrative boundary' were dropped, the overall pattern of internal movement (as measured from the census) would highlight the importance of local residential relocation and emphasize that longer-distance migration was only one part of a more general pattern of movement.

This is not to suggest that the methodology of demography is in need of a total rethink. There are positive advantages to the maintenance of a commonly accepted framework of analysis which allows for consistency over time and between different localities. If the terms of analysis were constantly redefined, it would introduce confusion about whether the identified changes in population resulted from actual developments or from variations in the way the data were represented. However, there are possibilities for new approaches in order to further the understanding of contemporary issues. The example of the application of a life-style classification rather than the Registrar General's socio-economic groups illustrates one way in which the methods of demography could keep pace with the changing structures within the British population.

# 4

# BRITAIN'S POPULATION HISTORY

An understanding of contemporary and future trends in Britain's population relies on knowledge of what has happened in the past. The analysis and interpretation of former events help to shed light on the relationship between demographic behaviour and economic and social change. Short-term developments need to be viewed in the context of longer-term patterns, with a clear understanding of the factors that have determined the rate and direction of population change. During the past two centuries there have been significant variations in the underlying trend of population change in Britain as a result of changes in mortality and fertility. These changes have had a profound impact on the perception of population issues and have influenced population forecasting and policymaking at different periods in the past.

## Population transition

The progress of population growth in Britain from the middle years of the eighteenth century until the 1930s represents a major demographic transformation. Prior to the onset of industrialization, population had generally been kept in check by the combined effect of relatively high levels of mortality and social and economic restrictions on fertility. Economic restructuring removed one of the major constraints to growth – dependency on resources – and ushered in a new era of rapid population expansion (Wrigley and Schofield 1981). Britain's population grew from approximately 6 million in 1750 to 42 million by 1931, during which period there were major fluctuations in both mortality and fertility (Figure 4.1). The crude death rate (for England and Wales) fell from approximately 27.5 per 1,000 in 1750 to 12.3 per 1,000 in 1911, and the crude birth rate fell from 34.8 to 15.8 per 1,000 (Wrigley and Schofield 1981, OPCS 1987a).

Traditionally this process of population growth has been referred to as 'demographic transition' and general principles have been derived from the experience of early industrializing countries like Britain which, it is argued, hold true for most countries experiencing economic development.

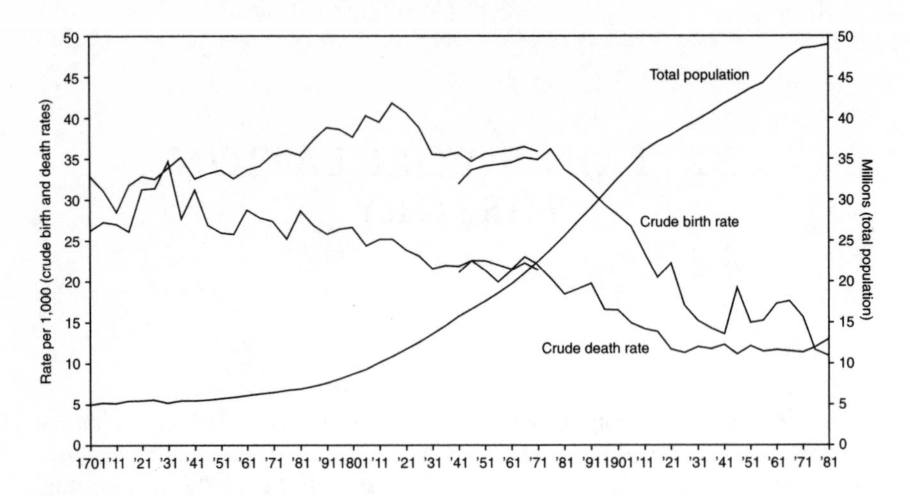


*Figure 4.1* Population change (England and Wales) 1701–1981

*Source:* Based on figures in Wrigley and Schofield (1981) 1701–1871, Mitchell and Deane (1962) 1837–1938, *Annual Abstract of Statistics* (1938–81)

*Note:* The discontinuity between 1873 and 1871 reflects the fact that in the early years of civil registration of births and deaths there was a significant undercount of events. The figures from Wrigley and Schofield are derived from estimated trends.

The demographic transition model is driven by economic change. First, improving living standards bring about a decline in mortality and initiate the process of growth. Later, with rising real incomes and other sociocultural developments, a trend is established towards later marriage and smaller family sizes and hence a reduction in general fertility. Eventually birth rates decline to a level close to death rates and a new stable population regime is established.

Considerable doubt has been cast on the value of demographic transition as an explanatory model. It is simplistic and places undue emphasis on the economic determination of population change. Nevertheless, it remains a useful framework for investigating the process of population growth in Britain for two reasons. First, it identifies the significance of the varying relationship between fertility and mortality during the process of population growth and second, it highlights the importance of fertility decline as the key turning point in the transition.

One other factor to be considered, which is often neglected in applying the transition model, is the potential influence of migration. In the British context, the net effect of migration may have contributed to the overall decline in the pace of population growth after 1850, but its role as a determinant of change was only identifiable in the later stages of transition and was of relatively minor importance in comparison to the influence of mortality and fertility. Detailed statistical evidence for immigration and emigration is lacking for the nineteenth century. Estimates derived from the difference between natural increase (excess of births over deaths) and actual increase suggest that Britain lost approximately 3,375,000 between 1871 and 1931 (Royal Commission on Population 1949). The outflow had the effect of depressing overall population growth by roughly 15 per cent during this period (not including the additional children born abroad). Migrants arriving in Britain came mainly from Ireland and other parts of Europe. Emigration was directed towards Australia, New Zealand, North and South America and the spreading frontiers of the British Empire. As many of those involved in migration were younger adults, the consistent loss through migration may have had an impact on overall levels of fertility, particularly during the peak years for emigration in the early twentieth century (Figure 4.2).

The transition in mortality, viewed in the longer term, was a relatively gradual and progressive process. The downward trend began in the mid-eighteenth century largely as a result of the amelioration of fatal epidemic diseases. Smallpox, typhus and many other diseases had had a major impact on mortality in the earlier part of the century but their importance had declined by the 1750s largely because of improving living standards and perhaps also the introduction of inoculation against smallpox. Such a downturn was not particularly remarkable – patterns of mortality in the past had been subject to regular ups and downs. What was significant was the fact that, apart from minor reversals, the improving conditions of life were sustained.

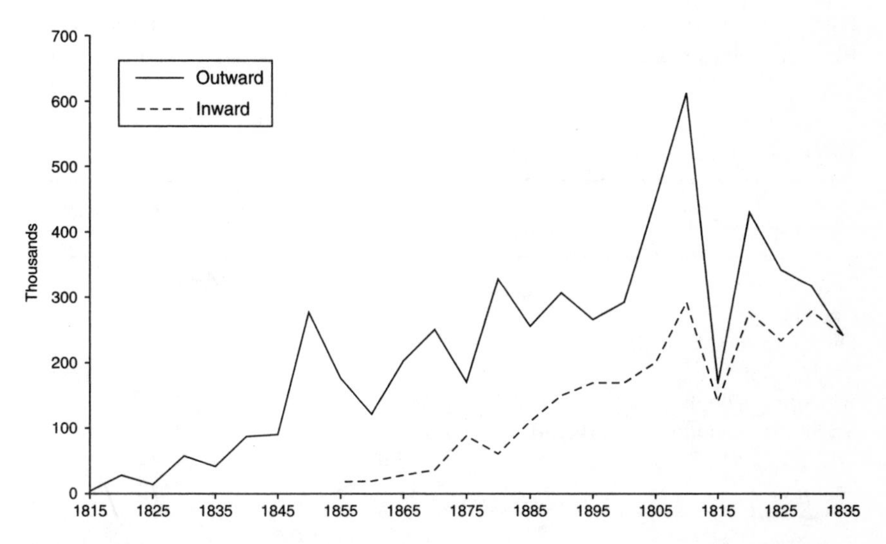


*Figure 4.2* Migration 1815–1935

*Source:* Mitchell and Deane (1962)

The general trend of improvement, however, was interrupted in the nineteenth century by the unhealthy conditions which developed in the rapidly expanding towns and cities. By 1850 over half of the population had become urban dwellers and the problems of insufficient accommodation and inadequate sanitation had encouraged the spread of epidemic diseases. The main killer diseases were typhus, cholera and tuberculosis, although many other infectious diseases such as scarlet fever, dysentery and infantile diarrhoea were all too common in urban areas (Table 4.1). The towns acted as repositories of disease, with the poor environmental conditions encouraging the development of infection and with a constant stream of migrants replenishing the population at risk and spreading contagion. Death rates were particularly high for infants and children. In Liverpool, for example, only half of those born during the 1840s could expect to reach the age of five (Chadwick 1842).

Disease control and the ultimate reduction in urban mortality came about as the result of a number of different factors. Intervention by both central and local government into the management of towns undoubtedly had some impact. The pioneering work of the sanitary reformers in the 1830s and 1840s resulted in local government by-laws and central government legislation (1848 Sanitary Reform Act) for improving basic conditions such as the disposal of sewage, the provision of fresh water and the paving of streets. But the real achievements came later in the nineteenth century with the clearance of poor-quality housing from the core areas of cities and the management of new housing development through building controls. Alternative explanations of mortality decline focus on the general rise in levels of nutrition and standards of life, on possible changes in the prevalence of micro-organisms, or on the improving organization of health care

*Table 4.1* Urban mortality by cause of death, selected areas

| | *London registration area 1849* | *Liverpool 1861–70* |
|---|---|---|
| Total deaths | 68,755 | 460,370 |
| Cholera | 14,125 | 4,255 |
| Diarrhoea (and dysentry) | 3,463 | 51,911 |
| Typhus | 2,482 | 9,297 |
| Whooping-cough | 2,349 | 32,551 |
| Scarlatina (including diphtheria) | 2,149 | 30,213 |
| Measles | 1,154 | 25,514 |
| Smallpox | 521 | 5,175 |
| Other | — | 12,093 |
| Total | 26,243 | 171,009 |

*Source:* Gale (1959: 136, 138)

(Woods and Woodward 1984). It is impossible to judge the contribution of each set of factors separately as clearly there were close links between them, although greater emphasis has been given to the general improvement in environmental conditions (including diet and living conditions) than to the role of medical science or the changing character of disease (McKeown 1976).

The transition in mortality can therefore be explained in terms of the changing circumstances of industrializing Britain. Whatever the reasons for the beginnings of improvement in the eighteenth century, the management and gradual elimination of disease can be seen as a beneficial consequence of development. The transition in fertility, which began towards the end of the nineteenth century, is less easily explained in terms of clear-cut causal factors. The birth rate had risen consistently in Britain during the period 1750 to 1820 as a consequence of the relaxation of social and economic constraints on fertility. Industrialization brought with it greater opportunities for employment and severed the link between agricultural productivity and population growth. An increase in nuptiality and a trend towards earlier marriage resulted in an increase in crude birth rates from 34.8 per 1,000 in 1750 to 41.9 per 1,000 in 1822 for England and Wales (Wrigley and Schofield 1981). During the middle years of the nineteenth century, fertility levels remained relatively constant. The introduction of birth registration in 1837 provides more detailed statistical evidence for the measurement of population change from the mid-nineteenth century onwards, although information for the earlier years is influenced by a probable under-registration of events. The general fertility rate (GFR) peaks at 154 births per 1,000 women aged 15–44 in the five-year period 1876–80. From then on fertility showed a marked downward trend, bottoming out in the period 1936–40 with a GFR of 61 per 1,000. This quite dramatic transformation in fertility reduced the rate of natural increase in England and Wales from 1.6 per cent in 1876 to 0.02 per cent in 1940. By the 1930s Britain's population had reached the end of demographic transition and had achieved a new stability.

Many factors have been advanced to account for this decline in fertility without providing a totally comprehensive explanation (Woods and Smith 1983, Coale and Watkins 1986, Woods 1987). The experience of Britain was by no means unique – the trend was common across Northern and Western Europe in the late nineteenth century (Figure 4.3). Consequently, it has been usual to account for fertility decline in terms of the general underlying processes of change within modernizing societies. Industrial capitalism encouraged smaller families in order to secure the fortunes of the family unit and to promote the interests of the individual. The Royal Commission on Population reported in 1949:

> In the individualistic competitive struggle, children became increasingly a handicap, and it paid to travel light. The number of children

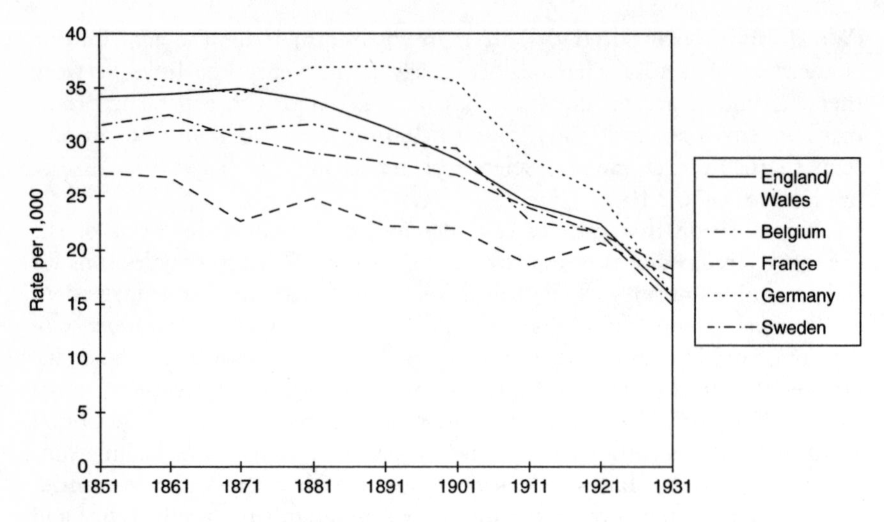


*Figure 4.3* Fertility transition in Europe 1851–1931

*Source:* Mitchell and Deane (1962)

> tended to be limited also, not merely because the expenditure upon them might handicap parents in maintaining their own standards of achieving their ambitions, but because the fewer children in the family the more could be spent on each child, and the better start it might have in life.
>
> (Royal Commission 1949: 39–40)

Other factors were also involved. The decline in infant and child mortality ensured that more children survived to adulthood and provided an incentive for limiting the number of pregnancies within marriage. The introduction of legislation and changes in the labour market effectively removed the earning potential of children within the family economy. At the same time the introduction of elementary education required parents to maintain children, at least until the age of ten. Also the onset of fertility decline coincided with the downturn in the British economy in the last quarter of the nineteenth century which squeezed general living standards and removed some of the optimism of the mid-Victorian boom years. In the same way the low levels of fertility achieved in the early 1930s were undoubtedly associated with the slow recovery of the economy during the 1920s and the direct effects of the depression during 1929–33.

These factors all help to outline the context within which fertility decline took place. The actual means by which it was achieved was mainly by the practice of family limitation within marriage, the net result of a general desire for smaller families. There was little change in the other parameters of fertility. The marriage rate remained relatively stable throughout the period (Figure 4.4) as did the mean age of marriage for women. The illegitimate birth rate fell but this only accounted for a small proportion of total births and its decline was in line with the trend in general fertility. The move to a small family norm can be linked to the interaction of a web of social, economic and cultural factors, all of which encouraged new patterns of behaviour and new attitudes towards family formation. The methods of restricting fertility became more widely known and practised after 1870, not specifically because of the introduction of new techniques or new forms of contraception, but simply the more general acceptance and application of existing methods. The virtues of family planning were publicized by a number of individuals and organizations, partly out of concern for the prevailing trends in population growth (e.g. the Malthusian League) and partly for more humanitarian motives to improve the quality of family life and remove the burden of unrestricted child-bearing. This publicity undoubtedly raised people's awareness of the benefits of contraception; it gave additional respectability to existing practice and eventually removed established taboos to intervention in the natural process of conception (Mitchison 1977). By the mid-1930s approximately two-thirds of women had used some method of birth control during married life, and of these roughly half had used appliance methods (Lewis-Fanning 1949).

The longer-term impact of these changes in family formation were significant and resulted in a sharp drop in the average size of families. For women married during the 1850s the average completed family size was between 5.5 and 6 children. For the cohort married during the period 1925–9, this figure had fallen dramatically to 2.2 children (Royal Commission on

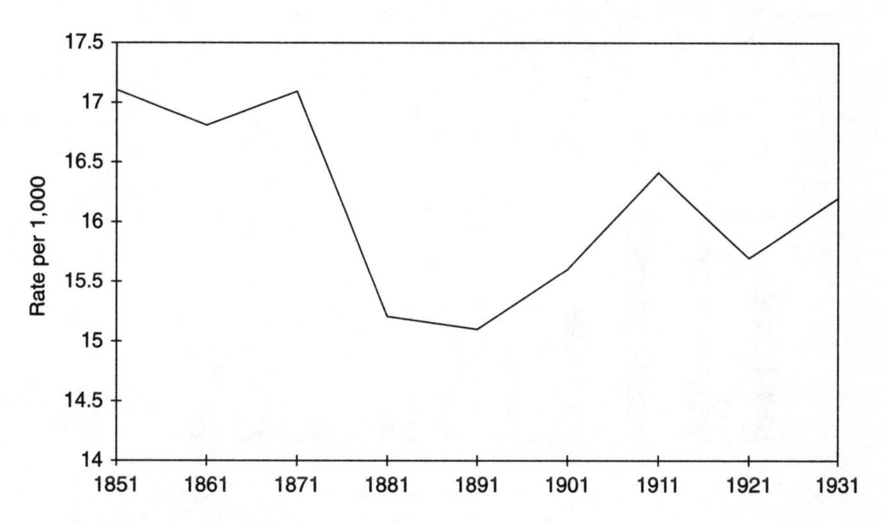


*Figure 4.4* Marriage rates 1851–1931

*Source:* OPCS (1987a), table 1.5

Population 1949). The decline in average family size was also accompanied by a reduction in the range of family sizes (Figure 4.5). In the mid-nineteenth century there was a fairly even distribution across the various sizes of family. Eighty per cent of families had more than two children and families of ten or more were not uncommon. By the later period this distribution had become very much more concentrated. Two-thirds of families had two children or less and only 4.3 per cent of families had more than six children. In addition, the proportion of childless couples had nearly doubled from 9 to 17 per cent of families.

Whilst this new pattern of smaller families had become well established by the 1930s, there were still significant variations between individual groups within British society. Some occupational groups had traditionally maintained higher levels of fertility, partly in response to income levels and partly because of the lack of employment opportunities for women (e.g. coalminers, agricultural workers and dock labourers). The decline in fertility was pioneered by the professional classes where smaller families were common before the general onset of fertility decline. This group had experienced a 33 per cent reduction in average family size between the cohorts marrying in the 1850s and the early 1880s (Royal Commission on Population 1949). By the turn of the century it was the non-manual wage-earners and skilled groups which were demonstrating the most rapid rates of decline. Family size in the unskilled category also declined markedly over the period, but having started from a higher level it took significantly longer for this group to achieve the small family norm. By the 1920s average family size was still 3.35 children per family, compared to 1.75 for professional groups (Glass and Grebenik 1954). There was approximately a twenty-year lag between the 'middle' class and the labouring population. Birth rates achieved

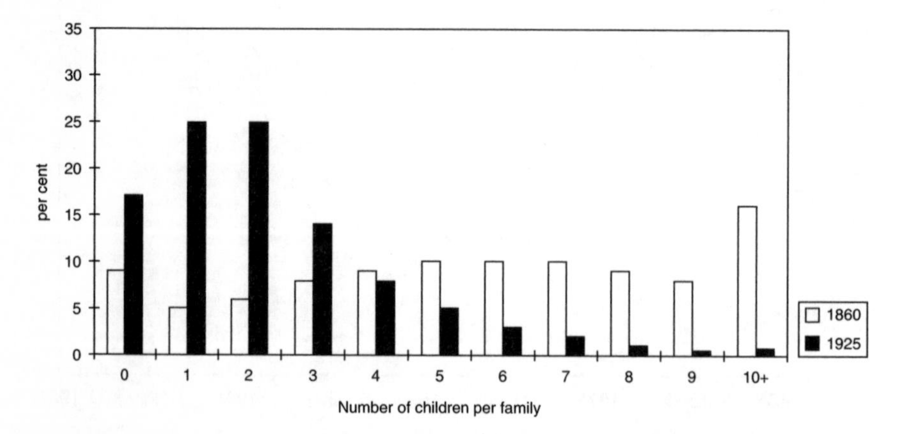


*Figure 4.5* Number of families by family size 1860 and 1925

*Source:* Royal Commission on Population (1949)

by skilled workers in 1911 (153 births per 1,000 husbands under 55 and wives under 45) were not attained by unskilled workers until 1931 (Hubback 1947: 28).

## Population issues in the 1930s and 1940s

The year 1933 marked the low point of fertility in Britain. The total period fertility rate of 1.72 is the second lowest ever recorded (the lowest was in 1977) and the net reproduction rate of 0.74 indicates that by this period the population had fallen below the level of natural replacement. Not enough children were being born to sustain the population at its current size and the longer-term trend pointed towards a gradual reduction in the future (OPCS 1987a). At the time this appeared to represent the normal state of affairs for the new demographic regime that had evolved over the past century. The same pattern was apparent in other Western European countries and there was little reason to expect that the trends in births and deaths would be reversed. The process of fertility decline had percolated down the social hierarchy and small families had become widely accepted. Further improvements in mortality might have been expected to off-set some of the decline in fertility, but by this time the death rate had already fallen to a relatively low level and any future changes were likely to be fairly gradual. For the first time commentators began to consider the implications of a 'no growth' scenario in population or, in some cases, the more worrying prospect of population decline and the 'ultimate threat of a fading-out of the British people' (statement by the Royal Commission on Population, September 1945, quoted in Hubback 1947: 14). These concerns were expressed by Enid Charles in *The Menace of Under-population: a Biological Study of the Decline of Population Growth* (1936), in which she claimed that:

> The prosperous classes of industrial nations, like other ruling castes in the past, have become the victims of their own ideology. In seeking to mitigate poverty by preventing the poor from reproducing they have moulded the destiny of a civilization which has lost the power to reproduce itself.
>
> (Charles 1936: 2)

Population projections, based on the assumption of constant birth and death rates at the 1939 level, resulted in a total population size of 33.54 million in 1999, falling to 25.20 million by 2039 (slightly more than half its 1939 level). This was seen as an immediate and worrying state of affairs. 'For all practical purposes we can regard the 1946 population as the largest Britain will have' (Abrahams 1945: 22). The report of the Royal Commission on Population in 1949 echoed the same basic prospect. Despite the increase in births in the later 1940s (which was put down to short-term post-war resurgence), the longer-term expectation was for a slight rise in population to the 1970s and then a long and gradual glide-path of decline into the twenty-first century. The implications of this prognosis were considered to be profound, not simply in terms of the total size of the nation's population but more specifically in relation to its age structure. The consequence of fertility levels remaining below replacement, coupled with improvements in mortality, was an ageing population. Britain faced the prospect of a smaller and older population in the second half of the twentieth century.

The effects of this shift in the age balance of the population were of considerable significance both in a contemporary context (as for example with the planning of educational provision) and more generally in the relationships between numbers in different age-groups, particularly between those of working age and the dependent population. The expected increase in the number of people over 65 (from 5 million in 1947 to 7.3 million in 1977) would more than double the cost of retirement pensions as well as substantially increase the demands on health and welfare services. Similarly, the decline in the numbers of children in the 1920s and 1930s would influence the size and structure of the labour force, creating an imbalance between younger and older members. A shortage of younger workers would make it more difficult for expanding industries to grow. Older workers were seen to be less mobile and more resistant to change. In addition, too many members of the labour force in the older age-categories would frustrate opportunities for promotion and would create a cause for discontent. Such changes would also alter the balance between producers and consumers in the population, with a possible impact on the size and purchasing power of the home market. The report of the Royal Commission on Population went further to suggest that such an upset to the age balance of the population could have more serious social effects,

> . . . a society in which the proportion of young people is diminishing will become dangerously unprogressive, falling behind other communities not only in technical efficiency and economic welfare but in intellectual and artistic achievement as well.
>
> (Royal Commission on Population 1949: 121)

These changes were also seen to have more wide-ranging implications for the security of the country and Britain's national status as a world power. In the changing geo-political structures of the 1940s, Britain's future position was unclear. Overshadowed by the larger populations (and military strength) of the Soviet Union and the US and aware of the potential for population growth in China and other developing countries, the consequences of population decline could mean that Britain would slip down the world rankings to become a country of limited significance:

> Britain, at least as much as any other great nation, has an immense part to play in the world; her standards of value – the democratic ideals of freedom, kindness, tolerance, justice and the rule of law – are of such fundamental importance to the welfare of the world that it is highly desirable that her present considerable share of influence should at least be maintained, if not increased. While it is true that mere numbers of heads do not in themselves enhance either influence or prestige – since this mainly depends on intelligence, energy, tradition and values – still numbers are a limiting factor. To fall much below the present level of population in this country would inevitably relegate us to the category of small nations and seriously reduce our influence, political and cultural. In countries like Sweden, Denmark, Holland or Belgium there is much to admire and a very high level of civilisation has been reached. But their influence on the world is negligible compared with that of Great Britain, whose culture is in most respects as high or higher and whose numbers are very much greater. Small countries can now exist only if permitted to do so by the larger ones.
>
> (Hubback 1947: 271)

This may be slightly overstating the case but what it illustrates is the profound impact that population issues can have on public consciousness and national pride. The debate about the future prospects of the nation's population was of significance not only because of the more general issues and long-term consequences, but also because of the more immediate influence it had over the formation of policy by central government. A mark of the Government's concern was the establishment of the Royal Commission on Population in 1944:

> . . . to examine the facts relating to the present population trends in Great Britain; to investigate the causes of these trends and to consider their probable consequences; to consider what measures, if any, should be taken in national interest to influence the future trend of population; and to make recommendations.
>
> (Royal Commission on Population 1949: 1)

The extensions to the Welfare State and other elements of social legislation which were included in the post-war reconstruction were, in part at least, a response to the perceived need to maintain the size of the labour force and encourage a return to replacement levels of fertility. The blueprint for welfare state legislation (the Beveridge plan) included, amongst a range of new benefits, a system of family allowances to be paid for each child after the first (introduced in 1945) and the provision of maternity grants (included in the National Insurance Act in 1946). The establishment of the National Health Service, the extension of local authority housing provision, the building of new towns and a whole host of other reforms introduced by the Labour Government provided a new supportive structure for families in Britain, particularly for the least well-off in society.

## Population trends 1945–70

The gloomy predictions of the 1930s and 1940s were proved wrong by events in the two decades after the Second World War. Fertility levels recovered from the low point of the mid-1930s and the underlying trend in the British population switched from long-term decline to potential growth. The immediate cause of this turn-round was the so-called 'baby boom' of the 1940s. The total period fertility rate for England and Wales increased from 1.78 (1931–5) to 2.39 (1946–50), and there was a marked upturn in the number of marriages. This was accompanied by a shift in age-specific fertility rates for women in the younger age categories (15–19 and 20–24). The total number of births reached 780,933 (1946–50), nearly a 30 per cent increase on the early 1930s (OPCS 1987a).

At the time there appeared to be nothing very remarkable about this event. There had been a similar peak in births in the years immediately following the First World War (Figure 4.1) and it was easily explained in terms of the circumstances of the time. The outbreak of war had encouraged a substantial increase in the number of marriages (482,000 in 1939–41, 21.5 per cent more than in 1936–8), and after the war there was a further period of high figures (441,000 in 1945–7). As a result the number of births had begun to rise during the war years and reached a peak in 1946–7 following the return to normality. The period in general was a time of recovery and reconstruction with more or less full employment and a feeling of optimism about the future – factors which might reasonably be expected to be conducive to family formation.

The Royal Commission on Population acknowledged the significance of this upturn, particularly its longer duration than the First World War peak and its possible origins in the later 1930s. However, they found no evidence to suggest that it marked a significant turning-point in population trends. It was largely accounted for by changes in the marriage pattern, with the war encouraging people to bring forward the age of marriage and family formation – a 'borrowing of marriages from the future' which created a situation that would be difficult to sustain in the longer term. Earlier and more universal marriage had the effect of concentrating births into a relatively short space of time rather than marking an absolute rise in the numbers of children that an individual cohort would produce. The expectation was that the increase in births would be balanced out by relatively lower fertility over the following decade and a return to stable population conditions (Royal Commission on Population 1949).

This indeed was the case in the early 1950s. The total period fertility rate dropped back to 2.18 and the net reproduction rate approached 1.0 (1951–5). The total number of marriages (and the crude marriage rate) also declined. Age-specific marital fertility remained relatively high for the younger age-categories but declined more significantly for women aged over 25. It appeared that although the trend towards younger marriage had been maintained, it had not resulted in significantly larger families (average of 2.3 in 1955). In comparison with the 1930s, there had been a decline in childless families and families with only one child. At the same time there had been an increase in two-child families (the modal class) and a smaller rise in families with three or more children. Seemingly this provided clear evidence of planned parenthood; families were simply being completed earlier in the life-cycle. After the disruptions of the 1940s, Britain's population was back on the rails, heading towards a stable population regime with little or no further growth in numbers. Population projections in 1955 for the year 2000 estimated a total population for England and Wales of 47,000,000, an increase of less than 6 per cent (2,559,000) over forty-five years, resulting mainly from a gradual decline in mortality.

In the event this projected figure was exceeded within ten years. After 1956 there was a relatively sudden and substantial increase in fertility which resulted in a short, sharp upturn in growth rates. The total period fertility rate rose progressively from 2.22 in 1955 to a peak of 2.93 in 1964 – a rise which represented a 30 per cent increase in the annual number of births and which added an additional 3 million to the population (Figure 4.1). This quite remarkable turn-about was totally unexpected and was all the more surprising as it occurred at a time when the cohort of women in their twenties was relatively small – a consequence of the lower levels of fertility during the 1930s.

Given the detailed statistical evidence available, it is possible to identify the main characteristics of this second 'baby boom'. It is not so easy, however, to provide a comprehensive explanation as to why it happened. As with the decline in fertility at the end of the nineteenth century, it was largely a reflection of changing behaviour and changing attitudes towards preferred family size and the timing of marriage and family formation. Age-specific fertility rates (Figure 4.6) indicate a marked increase in the number of births per 1,000 women across the age-range, but with the most significant increases occurring in the younger categories. In 1955 441,727 children had been born to women under 30 (65 per cent of the total). In 1964 the comparable figure was 626,083 (71 per cent). This age-shift in maternity was accompanied by a lowering in the average age at first marriage for women from 23.9 to 22.8. In 1965 33 per cent of women marrying were under 20 (Williams 1972). Over the same time period there was a 100 per cent increase in the number of illegitimate live births which by the 1960s made up a larger proportion of total births (7.2 per cent in

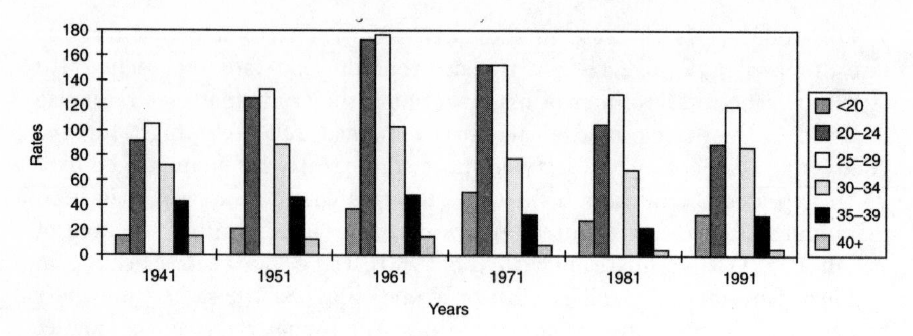


*Figure 4.6* Age-specific fertility 1941–91

*Sources:* OPCS (1987a), table 3.2, and *Population Trends* 88, table 9

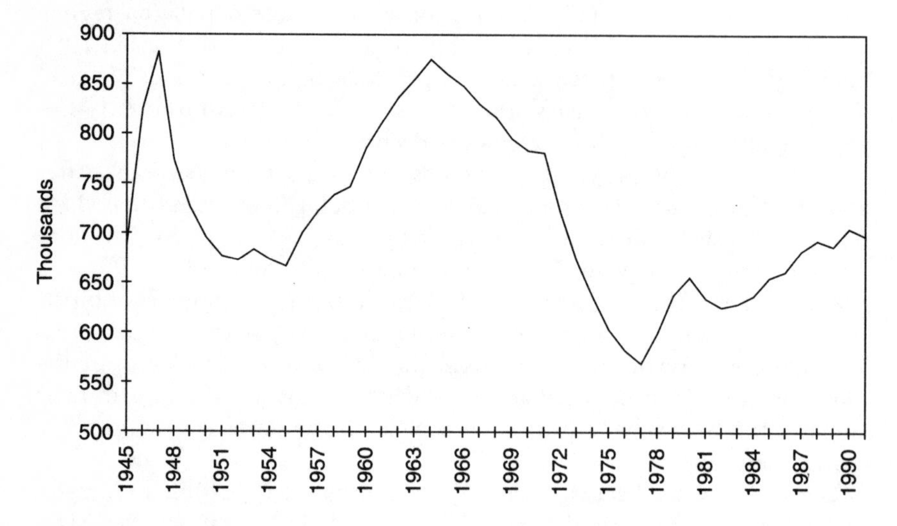


*Figure 4.7* Total births (England and Wales) 1945–91

*Sources:* OPCS (1987a), table 1.1, and *Population Trends* 88, table 8

1964). There was also an increase in the number of pre-marital conceptions (OPCS 1987a).

This relatively rapid turn-round in the prevailing trends in fertility can be partly accounted for by the varying patterns of behaviour of successive cohorts within the population. A shift in the age of maternity resulted in the coming together of peak periods of fertility for women in different age categories. The cohort of women who started child-bearing in 1951–5 were experiencing peak fertility between the ages of 25 and 29 at the same time as the following cohort of women, who commenced child-bearing five years later and who experienced peak fertility between 20 and 24 (Figure 4.6).

However, this still leaves the question as to why a particular cohort should prefer earlier marriage and child-bearing, and larger families? The quite dramatic change in demographic behaviour is partly a reflection of the social and economic context of the time. The generation that commenced child-bearing after 1955 was influenced by a different set of determinants to those which had prevailed in the immediate post-war years. Britain was experiencing a period of sustained economic growth and full employment, which for many had resulted in a significant improvement in the material standard of living. At the time of the 1959 general election the Conservative Government under Harold Macmillan claimed that 'You've never had it so good!' which may or may not have been true, but still reflected a general sense of improvement and optimism about the future. Underlying trends in the performance of the economy influenced the decisions taken by young adults about when to marry, when to begin a family and how many children to have, which collectively determined the course of fertility.

There may also have been less tangible factors, which are difficult to quantify, but nonetheless had an impact on contemporary trends. Changes in fashion and the spread of more liberal attitudes meant that the new generation of teenagers and young adults were less constrained by the traditional codes of behaviour that had determined the social life of their parents. Sexual relations before marriage became more commonplace, as did the cohabitation of unmarried partners. The stigma that had previously been attached to illegitimacy began to evaporate. The so-called 'sexual revolution' of the 1960s may not have been the root cause of the upturn in fertility (which was still largely confined within marriage) but it contributed to the changing social context of the period.

Collectively these various factors helped to establish a new demographic regime which in many ways represented a reversal of previous trends. What had been taken for granted as persistent features of Britain's population no longer applied. As Jean Thompson, the Chief Statistician at the General Register Office, noted:

> One feature of the post-war years worth particular notice has been the emergence of the professional classes, the highly educated, those who occupy positions of considerable influence in society as being a group of relatively high fertility. These are the kind of people who are, and were previously, those most capable of achieving just the family size they desire. Their behaviour can be quoted as one illustration of the general thesis that the resurgence of fertility in the post-war years has shown many features that lead to the conclusion that it must largely have reflected a change in the size of family people have both desired and felt capable of supporting.
>
> (Taylor 1970: 12)

Whereas previously it had been assumed that fertility was inversely related to social class (the higher the status the smaller the family) and that general improvements in material well-being would eventually bring about the universal adoption of small families and a stable population, now it appeared that the opposite might be the case. Higher standards of living could lead to larger families. Demographic transition had bottomed-out and was beginning to go into reverse!

So far most attention has been given to the varying pattern of fertility during the post-war period, as this had the most immediate impact on the pace and direction of population change. Variations in mortality and migration were less significant in terms of their contribution to overall growth, although both experienced some important developments during this period. As indicated earlier, the decline in mortality during the twentieth century was a gradual and continuous process. There was little or no change in the crude death rate between the late 1940s (11.5 per 1,000) and the late 1960s (11.7 per 1,000). But this is mainly a reflection of the general ageing of the population rather than a stable pattern of mortality. Figures for life-expectancy from birth show an increase for males from 66.2 years (1951) to 68.6 years (1971) and for females from 71.2 to 74.9. Much of this apparent extension to life was accounted for by reductions in infant mortality, particularly in the very early stages. Both neonatal (deaths in the first four weeks) and perinatal (still-births and deaths under one week) figures declined sharply (Table 4.2) – an achievement which was largely the result of developments in medical care and improved maternity facilities.

For the population as a whole the most notable feature was the continued decline in importance of infectious diseases. This period effectively marks the end of the process of disease control. The major killer diseases had been on the retreat since the middle of the nineteenth century, mainly due to improving conditions of life (McKeown 1976). In the years after the Second World War the implementation of universal immunization against diseases

*Table 4.2* Infant mortality (UK) 1946–71

| | *Still-births*[a] *(per 1,000 live births)* | *Perinatal*[a] | *Neonatal* | *Post-neonatal* |
|---|---|---|---|---|
| *1946* | 9.0 | 9.3 | 6.7 | 17.8 |
| *1951* | 7.8 | 8.2 | 3.5 | 11.7 |
| *1956* | 7.6 | 6.9 | 2.7 | 7.2 |
| *1961* | 7.9 | 5.8 | 2.1 | 6.4 |
| *1966* | 6.8 | 4.7 | 1.8 | 6.4 |
| *1971* | 6.1 | 4.0 | 1.8 | 6.0 |

*Source: Annual Abstract of Statistics* (various)

*Note: a* includes still-births

such as tuberculosis, diphtheria and measles completed the process. By 1971 only approximately 13 per cent of all deaths were the result of infectious diseases and almost all of these were accounted for by bronchitis and pneumonia contracted by the elderly. As a consequence, there was a shift in emphasis towards the degenerative diseases of old age. Deaths from cancer, ischaemic heart disease and other circulatory conditions all increased significantly during this period and represented over 70 per cent of deaths by 1971 (Table 4.3).

These changes in mortality only had a marginal impact on population growth. In a similar way the net effect of migration between 1945 and 1970 made little contribution to the general trends. Overall Britain lost marginally more people through emigration than it gained from immigration, although the patterns of inflow and outflow varied considerably during the period. The main destinations for emigrants were still the Old Commonwealth countries (Australia, New Zealand and Canada), South Africa and the US. Levels of emigration peaked during the 1950s when the Old Commonwealth countries were actively recruiting skilled and professional labour from Britain. Immigration, on the other hand, built up progressively in the years after the war and reached a peak in 1961 of 500,000 (Figure 4.8). In the late 1940s many of these migrants came from other parts of Europe, displaced by the effects of war and responding to labour shortages within the British economy. The Government played an active role in encouraging immigration from the Baltic states, Poland, Ukraine, Italy, Austria and Germany (Holmes 1982). By the mid-1950s this stream of migration was overtaken by much larger numbers arriving from various parts of the New Commonwealth, particularly from the West Indies and the Indian sub-continent. West Indian migration had begun during the war but gathered pace during the early 1950s, partly in response to immigration restrictions in the US and partly as a result of active recruitment of West Indian labour by organizations such as the London Transport Executive, the British Hotels and Restaurants Association

*Table 4.3* Cause of death 1946–71, selected causes (England and Wales)

| | *All deaths* | *Cancer* | *Circulatory system* | *Respiratory diseases* | *Infectious diseases* |
|---|---|---|---|---|---|
| *1946* | 481,274 | 75,407 | 205,159 | 64,548 | 27,629 |
| *1951* | 549,380 | 86,080 | 222,027 | 75,290 | 17,048 |
| *1956* | 521,331 | 92,710 | 222,636 | 54,667 | 7,313 |
| *1961* | 551,752 | 99,915 | 230,502 | 67,740 | 4,737 |
| *1966* | 563,624 | 108,158 | 237,207 | 71,045 | 3,509 |
| *1971* | 567,262 | 116,895 | 293,546 | 74,562 | 3,300 |

*Source: Annual Abstract of Statistics* (various)

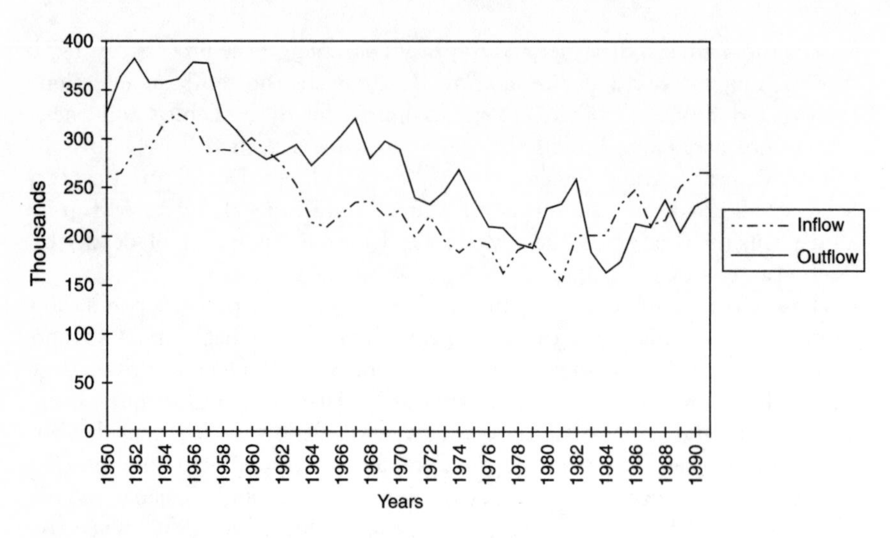


*Figure 4.8* International migration (UK) 1950–91

*Source: Annual Abstract of Statistics* (1951–94)

and the National Health Service. By 1961 the West Indian-born population had reached 172,000 and was concentrated mainly in London. Migration from India and Pakistan began later and was different in character. In the early stages immigrants were predominantly young adult males in search of employment opportunities in London and the manufacturing districts (notably the West Midlands and West Yorkshire). They were followed a few years later by a secondary migratory wave of dependent wives and children.

One reason for this two-stage pattern was the imposition of immigration restrictions in 1962. The steeply rising trend in New Commonwealth immigration had generated concern about the longer-term consequences of unrestricted entry, both in terms of the impact on total numbers (at a time of population growth) and more specifically about the problems of assimilating rapidly expanding migrant groups in British cities. The Commonwealth Immigrants Act effectively closed the door. Only the close relatives and dependents of previously settled migrants were permitted entry, and these were further restricted by subsequent acts in 1968 and 1971.

As a consequence, New Commonwealth immigration declined dramatically after the mid-1960s and the former pattern of emigration exceeding immigration was re-established. There were, however, some lasting effects from the peak years of immigration. The greater majority of those involved were adults in the reproductive age-band, which created a very distorted age/sex structure for the migrant communities. In addition, levels of fertility amongst recent migrants were significantly higher than the rest of the population. During the 1960s this group accounted for about 40,000 births a year, approximately 5 per cent of all births in Britain, although they made up only 2 per cent of the total population (Population Panel 1973). This was a temporary situation which changed during later years as the cohort aged and the fertility patterns of the group (and their subsequent offspring) moved towards the national average.

## Population issues during the 1960s

> The essential thesis . . . is that the British population is too large, is continuing to grow, and ought to be reduced to a much more tolerable level.
>
> (Brooks 1973: 159)

The changing course of population growth in the years after 1955 had brought about a total turn-about in attitudes to population issues in Britain. No longer were people concerned about the consequences of labour shortage, the burden of dependent pensioners or Britain's position in world population rankings. Instead rapid population growth presented a whole new array of problems related to the consumption of resources, the planning of social provision and a fear that the country may be exceeding its natural 'capacity' to maintain the population at acceptably high standards of living.

Many of these concerns were expressed by Edwin Brooks in *This Crowded Kingdom: An Essay on Population Pressure in Great Britain* (1973). Brooks was both an academic and a member of the Labour Government between 1966 and 1970. He was responsible for introducing the National Health Service (Family Planning) Bill in 1967 and was closely involved in the debates of the late 1960s about the problems of accommodating Britain's expanding population. His analysis of the situation was based on two primary concerns. First, that the country's population growth needed to be set in a global context. At a time of rapid world population growth and expanding demands on basic resources, population increase in the developed world (with its high levels of resource consumption and pollution) would simply exacerbate the problems of development in the Third World and further contribute to the destabilization of the world economic system.

Second, he expressed the view that increasing population pressure would result in the destruction of the country's natural resources and lead inevitably to a deterioration in the quality of life for British citizens. Over-population was identified with the relentless spread of built-up areas and the urbanization of the countryside, representing the 'tidal erosion of village culture' and leading towards the creation of a 'single urban region, a "megalopolis" or super-conurbation' (Brooks 1973: 57). Too many people would also create physical pressures of crowding and interpersonal friction which could result in stress and discontent. 'A larger population will most certainly guarantee greater frustration and unhappiness' and could 'unleash violent and perverted behaviour' (p. 162).

Rather like the gloomy predictions of writers in the 1930s, such comment needs to be interpreted in the context of the period. At the time there was much concern about world population explosion and the consequent shortages of food and other basic resources. These anxieties had been translated into doom-laden predictions about the future predicament of mankind, by publications such as Paul Ehrlich's *The Population Bomb* (1971) and the study commissioned by a group of wealthy businessmen (the Club of Rome), *The Limits to Growth* (1972). Whilst not everyone shared their view that drastic action was required to avert crisis, nevertheless it was generally acknowledged that continued population growth presented problems at both the global and national level.

In Britain the implications of fertility remaining at the levels achieved in the early 1960s were estimated as an additional 13 million people by the end of the century. At the time it was difficult to predict what impact these additional numbers would have on economic performance and the standard of living in the country. The House of Commons Select Committee on Science and Technology reported on the population of the UK during the session 1970–71. Their conclusion was that 'the Government must act to prevent the consequences of population growth becoming intolerable for the everyday conditions of life'. As a result the Government appointed a mixed panel of experts to review the current situation and make recommendations about the need for further action. Having considered a wide range of factors, the Population Panel reported in typically restrained language that 'on a wide variety of assumptions we are likely to be better off in terms of average material wealth in the year 2011 if population grows slowly than if it continues with the fertility rates of 1971, and worse off if we return to the high fertility of a few years ago' (Population Panel 1973: 78). They considered that the situation was not sufficiently worrying to require 'immediate policy initiatives designed to reduce dramatically the rate of increase'.

However, they did recommend that the Government should adopt 'an attitude to population' which would encourage a slower rate of increase and aim towards the goal of a sustainable stationary population in the future. They argued that 'policies towards health and welfare, social security, family allowances, housing, education and the distribution of tax burdens as between families of different sizes may all affect, directly or indirectly, people's decisions about the number of children they wish to have' (Population Panel 1973: 103). They also advocated an extension of family planning services to reduce further the number of unplanned pregnancies. What was needed was not a population policy but a concern about population in the formulation of social policy. They stopped short of proposing more persuasive measures to increase the cost of bringing up children, although this suggestion had been made earlier by other commentators. Milton Freeman (in a comment reminiscent of Thomas Malthus's views on the Old Poor Law) stated, at a conference on the Optimum Population for Britain, that:

> It is unrealistic . . . to promote such causes as celibacy, delayed marriages with chastity outside of marriage, or abstinence inside of marriage as means of achieving reduced fertility. It would be more expeditious to alter existing fiscal arrangements, so that, e.g. illegitimate or supernumerary offspring are heavily taxed, and tax exemptions apply to single persons or couples with one or no children (unless adopted), and higher pensions are paid to senior citizens without supportive children. At the present moment most advanced nations encourage high fertility through their tax structures and family allowance and welfare systems.
>
> (Freeman 1970: 147)

Although the Government recognized that there was no need to adopt drastic measures to control unrestrained growth, there remained the problem of accommodating the additional numbers that would be added to the population during the remainder of the century. As the title of an article by Richard Lawton stated, 'England must find room for more' (Lawton 1974). Projections by the OPCS estimated that the population would increase by 11 million (20 per cent) between 1969 and 2001, and that this increase had implications that would affect every aspect of life. 'The physical planning of this growth in numbers, and the infra-structure of services needed – housing, the transport network, education facilities, health services – pose some of the most fundamental problems that Government (national and local) and the community have to consider' (Thompson 1971: 24–5).

Some of these problems were more pressing than others. The bulge in births had created demands for additional places in primary schools within five years. This in turn placed demands on provision for secondary education and eventually on places in further and higher education. More capacity in the educational system required more fixed capital investment in the building of new establishments and the expansion of existing facilities; it also required many more staff. The Government responded to the shortage of teachers by increasing existing provision for teacher education and by developing teacher training colleges. This led to demands for more experienced staff to teach the teachers, and so the process continued. The reverberations of these policy developments in education were felt throughout the 1970s and 1980s as the tidal wave of children passed through the education cycle and under-provision was followed by over-provision in the various stages of the system.

Planning for developments in many other areas of social provision was based on similar assumptions of continued population growth. At the time there seemed little possibility of a return to a lower fertility regime and the bulge generation born between 1955 and 1964 indicated a further rise in births in the late 1970s, when they reached reproductive age. The changing attitudes of the 1960s had created a new set of demographic parameters which would influence the likely course of population change in the future:

> There are important elements in the patterns of marriage and family formation which over recent years appear to have been relatively stable: that is, there is no clear evidence at present either of motives in society in general, nor specific causal agents, to demonstrate the likelihood of significant future change. An example of this is the relatively high marriage propensity and young age at marriage which has been the dominant pattern in the post-war period: there is no evidence of a reversal of this, with a change to later marriage.
> (Thompson 1971: 22)

## Population trends 1970–90

One reason why the Population Panel had not come up with more radical proposals for government policy on population was the realization that births had declined quite sharply in the early 1970s. Although they did not place too much significance on this 'short-term movement', they nevertheless explored the possibility of fertility declining to a level which would off-set the latent potential for population growth in the population structure. In a perceptive final paragraph they noted that 'a movement towards an average family size consistent with a stationary population is likely to be associated with an increased demand by married women for employment opportunities and other opportunities for social participation outside the immediate family circle' (Population Panel 1973: 122).

However, they did not anticipate the speed with which fertility would collapse over the following five years. The number of births in Britain fell from 870,000 in 1971 (980,000 in 1964) to 632,000 in 1977 (Figure 4.1). This decline was as dramatic and unexpected as the increase had been in the years after 1955. The total period fertility rate reached an unprecedented 1.66 in 1977 and the net reproduction rate for the period approached the levels of the early 1930s. In both 1976 and 1977 the number of deaths marginally exceeded the number of births and, for the first time since detailed statistical records had been kept, the country experienced a net loss of population through natural change. In the following years there was a slight recovery, but throughout the early 1980s the number of births remained at around 700,000 and deaths averaged 650,000 – producing an annual percentage growth rate (taking account of migration) of approximately 0.1 per cent; effectively a stationary situation. Between 1976 and 1986 Britain's population increased by only 503,000.

As with previous changes in fertility, the causes were largely to do with changing patterns of behaviour and attitudes towards marriage and family formation. The most notable feature was the change in age-specific fertility rates (Figure 4.6), which indicated a marked shift in child-bearing towards the later periods of the reproductive age-band. For England and Wales between 1971 and 1986, the mean age of mothers at birth increased from 26.2 to 27.0 years (and for first births from 24.0 to 26.2 years). In 1971 47.0 per cent of children had been born to mothers under 25; by 1986 this figure had fallen to 37.7 per cent.

Other statistical indicators shed more light on the factors which contributed to the decline in fertility. The total number of marriages in Britain fell from a maximum in 1972 (468,380) to a minimum in 1982 (377,108) a drop of approximately 20 per cent. If remarriages are excluded, the decline was even steeper (29 per cent). The average age at marriage for women in England and Wales rose from 22.6 in 1971 to 23.6 in 1984. Over the same period the percentage of marriages of women under 20 fell from 31.1 to 17.9 (and to 10.2 per cent by 1989). Associated with this shift towards later marriage was a change in the distribution of family sizes. For women aged 40 in 1984 (towards the end of their reproductive years) 41.4 per cent had two children (as opposed to 32.2 per cent in 1974), 13.5 per cent had only one child (15.9 per cent in 1974) and 34.1 per cent had three or more (40.1 per cent in 1974). The number of women without any children remained more or less constant at about 12 per cent (Figure 4.9). However, the evidence from younger cohorts indicates that the proportion of childless families was increasing within the population as a whole (OPCS 1987b). The estimated distribution of family sizes based on the age-order specific birth rates for 1983 shows that there was a slight increase in childlessness and larger families in the early 1980s (Figure 4.9).

The decline in the number of marriages, and the increasing age of marriage, was partly accounted for by an increase in cohabitation. During the 1970s the proportion of married couples, who had lived together before marriage, more than doubled and by 1979 10 per cent of all unmarried women (aged 18–49) were cohabiting (OPCS 1987b). Cohabitation was partly responsible for a significant increase in the number of births outside marriage (8.4 per cent of all births in 1971, 21.4 per cent by 1986), although this also reflected an marked increase in single-parent families. A sharp upturn in the number of divorces, particularly in the early 1970s following the Divorce Reform Act of 1969, was one of the factors which contributed to a later average age of marriage (through remarriage) and to increased single parenthood and cohabitation. There was also a significant increase in the number of abortions during the 1970s, largely as a

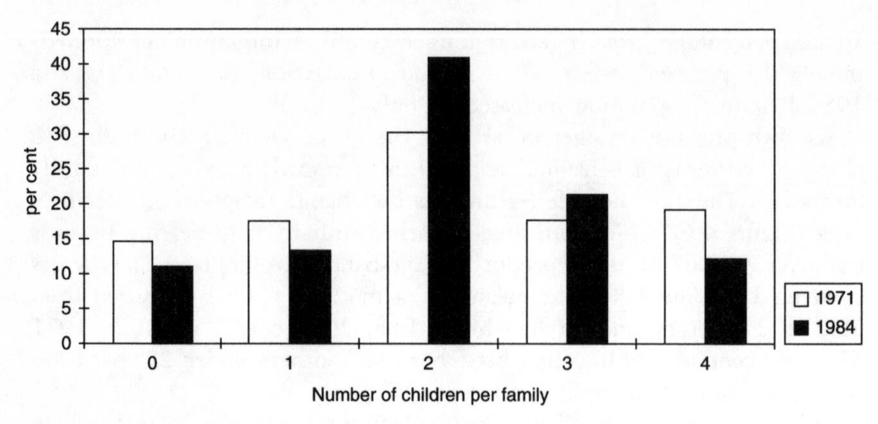


*Figure 4.9* Family size for women aged 40 1971 and 1984

*Source:* OPCS (1987b), table 4.20

consequence of the Abortion Act of 1967 which greatly widened the availability of legal abortion. By the 1980s family structures had become more complex, with a relative decline in the dominance of the nuclear unit and a rise in other family types (single parents, separated parents, regrouped families from divorced partners, etc.). One other notable feature of the changing patterns of family formation during this period was the differentials between social groups. For women under 25 the familiar pattern of higher fertility in the lower social groups was maintained, but for older women this situation was reversed. It was predominantly the professional and non-manual groups that had adopted patterns of later marriage and later child-bearing.

The explanation for the decline in fertility is closely related to the changing economic circumstances in Britain during the 1970s and 1980s. The impact of recessions in the mid-1970s and early 1980s resulted in significant changes to the structure of the British economy, notably a shift away from traditional patterns of heavy industry and manufacturing towards more capital-intensive methods of production and service sector employment. One of the consequences of this was a sharp increase in levels of unemployment which brought to an end the period of near full employment which had been sustained in Britain since the end of the 1940s. This, coupled with the effects of pay restraint and inflation, produced a relative decline in material standards of living. The worsening economic prospects for many people, particularly those living in the declining industrial districts who were dependent on semi-skilled and unskilled employment, encouraged the delay of marriage and family formation and effectively placed restrictions on family size. The climate of declining economic performance and restricted opportunity also had a depressing effect on family building in general, encouraging many to put off child-birth until conditions improved.

However, economic conditions only provide a partial explanation for the reversal in trends. The downturn in fertility had begun during the late 1960s at a time of relative economic buoyancy and long before the full effects of recession had been widely felt. Also the relatively brief period of recovery in the mid-1980s (at least in the South of Britain) did not result in any significant upturn in fertility. Other factors were also relevant. The rapid spread of oral contraception for women from the mid-1960s onwards, the extension of the Family Planning Service (following the recommendations of the Population Panel) and the wider availability of abortion, all provided a greater degree of individual control on fertility – both inside and outside of marriage. This resulted in a decline in so-called 'unwanted pregnancies' and gave parents the opportunity to plan both the timing and size of their families in relation to income and expenditure.

As predicted by the Population Panel, the changing employment opportunities for women were also an important factor. In 1971 56.7 per cent of women of working age were in the civilian labour force; by 1986 this figure had risen to 67.4 per cent. In absolute terms, the number of women working increased by about 2 million (21.3 per cent) and accounted for almost all the growth in the labour force as a whole during this period. A large proportion of these additional jobs were part-time (four out of five part-time jobs were occupied by women in 1988) but there was also a significant increase in career opportunities for women. By delaying marriage and the start of family formation, women were able to establish themselves in the labour force before taking time out for a relatively brief period of child-rearing (partly supported by the statutory entitlement to maternity leave). Although the interruption still presented difficulties for some women on re-entry to employment, and influenced the promotion and career development prospects for others, it nevertheless allowed many to combine successfully employment and family building. It also meant that couples were able to sustain the benefit of a dual income and thereby maintain relatively high standards of living. In these terms the 'cost' of having children was significantly increased and decisions about desired family size and the timing of child-bearing were strongly influenced by the sacrifice required in lost earning potential.

Related to this was the significance of changes in the housing market during the 1970s and 1980s. Over the period there was a marked shift towards home ownership and away from rented accommodation. At the same time rapid house-price inflation raised the cost of house buying in relation to average incomes. Consequently, a larger share of family income was diverted into mortgage repayment, raising the relative cost of family building. The situation became particularly acute in the South East of England during the mid-1980s when house prices rose very steeply and many women found themselves in a position of having to remain in employment to maintain the dual income necessary to cover housing costs. This may partly explain why there was no upturn in fertility of any significance in the South East in the mid-1980s – a period of expansion and relative prosperity in the region (see Chapter 6).

Changes in mortality and migration had only a limited impact on general trends in population over this period. The gradual improvements in mortality identified earlier were maintained, particularly for very young children. However, there was no real progress towards erasing the differentials between social groups at this end of the age-range and Britain's position regarding infant mortality did not improve as significantly as some other developed countries. There was an overall fall in mortality rates for those over 65 in the population, largely accounted for by a decline in deaths from diseases of the circulatory system (OPCS 1987b). Expectation of life improved by approximately 1.5 years for both men and women from the age of 60 (1971–86). The one exception to the general trend in mortality was the spread of AIDS as a killer disease, particularly for adult males, although the total number of deaths during the 1980s was relatively small (see Chapter 5). The pattern of international migration fluctuated from year to year (Figure 4.8). A net loss during the 1970s was replaced by a net gain during the 1980s, although the figures involved were relatively small (e.g. a net inflow of only 44,000 in the peak year of 1989). The direction of movement has shown some variation, with a continued decline in inflow from the New Commonwealth and an increased level of movement within the European Union.

Viewed in the longer term, it is possible to identify some general underlying causes to the variations in population change. Although at particular points in time the prospects for future change have looked worrying, it is clear that the ups and downs have been relatively minor fluctuations in an otherwise fairly stable population regime. Much of the short-term variation in fertility, for example, has been caused by differences in the timing of family formation by different cohorts and the overall impact on completed family size, viewed in retrospect, has been less dramatic. There can be little doubt that these fluctuations are related to longer-term changes in the performance of the British economy. The progression of boom and slump has varied the prospects for marriage and family formation in different time periods, creating distortions in the size of generations which have continued to reverberate down the years. These cyclical patterns will be of significance in the future in terms of the structure of the British population and the need for social provision. For example, the bulge generation born in the period 1955–64 will eventually reach retirement age in the late 2020s and create additional demands for financial support and health care.

The lesson from the past is that it is unwise to make predictions about the future on the basis of contemporary trends. However, the changes in the economy and society that have occurred over the past twenty-five years have strongly influenced demographic behaviour and imply that there is little likelihood of a return to higher levels of fertility in the foreseeable future. The total period fertility rate has remained below replacement level since 1973 and has maintained a constant level at around 1.8 for the past ten years. The recession of the early 1990s has reinforced the pressures on couples to delay child-bearing and restrict family size. Once more Britain is facing the scenario of a declining population, the problems discussed in the 1940s about a lack of young people and a surfeit of the aged are again issues of concern (see Chapter 7).

# 5

# RECENT TRENDS IN FERTILITY AND MORTALITY

Current trends in fertility and mortality partly reflect developments in the past as a function of longer-term trends. But the changing social and economic conditions in Britain also influence the context of population dynamics. Fertility is not an independent variable, unaffected by events, but is determined by the conditions of people's lives. For example, the increasing participation of women in the labour market has had a direct bearing on patterns of family formation and trends in fertility in recent years. Similarly, quality of life and life-style are determined by both social and economic factors. Patterns of mortality have a clear social dimension. Occupation, wealth and geographical location are all factors which determine the incidence of ill-health and life-expectancy. The population dynamics of Britain in the 1990s both reflect these underlying factors and in turn help to shape current social trends.

## Family formation

The low and relatively stable level of fertility of the 1980s has been maintained in the 1990s. The total period fertility rate for England and Wales in 1992 was 1.80 (falling to 1.75 in the final quarter). As identified in the previous chapter, there are essentially two factors which influence this prevailing trend in fertility. One is the variation in the size of generations and the age distribution of women within the population. The second is the behavioural patterns of the population and the changing attitudes towards marriage, family formation and desired family size. The former is entirely predictable given the consistency in mortality trends and can be determined from previous events. The latter has proved to be very much more difficult to explain or forecast. Over the recent past there has been a significant shift in the socially accepted norms of sexual behaviour and family structure which, for many parents and children, has resulted in a markedly different experience of family life to that of their parents and grandparents.

Family formation, whether within marriage or within a stable cohabiting relationship, generally results from rational decision-making by parents. Of course, not all conceptions are planned well in advance, but the universal availability of contraception and abortion allows adults to decide when to begin a family and the number and timing of children. These decisions are the key determinants of the prevailing trends in fertility. For any individual couple, a wide range of specific concerns may be considered before embarking on the process of family formation, but in aggregate terms the general trends identifiable in the 1990s can be associated with a number of related factors, including the declining significance and stability of marriage, the changing patterns of child-birth within marriage, the increasing financial 'costs' of a family and the varying social and economic context of family life.

## Marriage and divorce

It is no longer the case that the event of marriage marks the beginning of sexual relations and the start of family formation. In comparison to previous generations young adults are now less likely to marry and, if they do, they marry at a later age. After a significant fall during the 1970s, the marriage rate (number marrying per 1,000 persons aged 16 and over) remained relatively stable during the 1980s, but has fallen again since 1989 and is now below 40 per 1,000 unmarried persons aged 16 and over. The median age at first marriage for men has risen to 26.5 and for women to 24.6 (Figure 5.1).

*Figure 5.1* Median age at first marriage, mean age of divorce and remarriage rate 1971–91

*Source: Population Trends* 88, tables 21, 22 and 23

In contrast there has been a significant increase in the number of divorces. The divorce rate (divorce decrees per 1,000 married population) for England and Wales is now approaching 14 per 1,000 as compared with 11.5 per 1,000 in 1981. If this trend continues, roughly 40 per cent of current marriages will end in divorce (Haskey 1989). Over the same period there has been a decline in the median duration of marriage from 10.1 to 9.8 years (OPCS 1993c). The increase in divorce is consistent across the age-range of the married population and the mean age of divorce has remained relatively constant over the past ten years. Despite the upward trend in divorce, the number of divorced persons remarrying has remained relatively stable at around 80,000 per year in England and Wales. In relative terms this has meant a decline in the rate of remarriage (per 1,000 divorced persons aged 16 and over), particularly for those divorced at a relatively early age (Figure 5.1).

As a result of these trends in marriage and divorce, the proportion of both men and women aged 16–44 who are single has increased (Table 5.1). Between 1981 and 1991 the proportion of non-married males rose from 34 to 39 per cent and females from 39 to 44 per cent (Figure 5.2). Much of this apparent increase is accounted for by cohabitation of adults either as prelude to, or as a substitute for, legal marriage. The General Household Survey indicates that over a quarter of all men and women in their twenties were cohabiting in 1990, many of whom had been previously married (OPCS 1993d). Clearly the declining trend in remarriage for divorced persons has been matched by an increase in cohabitation, particularly for men. As much as 57 per cent of divorced men aged 25–34 were cohabiting in 1990, as opposed to just over a third of divorced women in the same age-group. Many more women remain single or as lone parents with the responsibility for children.

The increase in lone-parent families over recent years has been one of the most significant features of the changing patterns of family life and a topic of political debate in the early 1990s (see Chapter 7). In 1981 there were

*Table 5.1* Marital status 1981 and 1991 (England and Wales)

| | *1981* | | *1991* | |
|---|---|---|---|---|
| | *Males* | *Females* | *Males* | *Females* |
| *(all ages 16+)* | *(thousands)* | | | |
| Single | 10,614 | 9,424 | 11,371 | 9,816 |
| Married | 12,238 | 12,284 | 11,718 | 11,867 |
| Divorced | 698 | 2,939 | 1,221 | 1,497 |
| Widowed | 611 | 828 | 685 | 2,925 |

*Source: Annual Abstract of Statistics* (1997), table 2.4

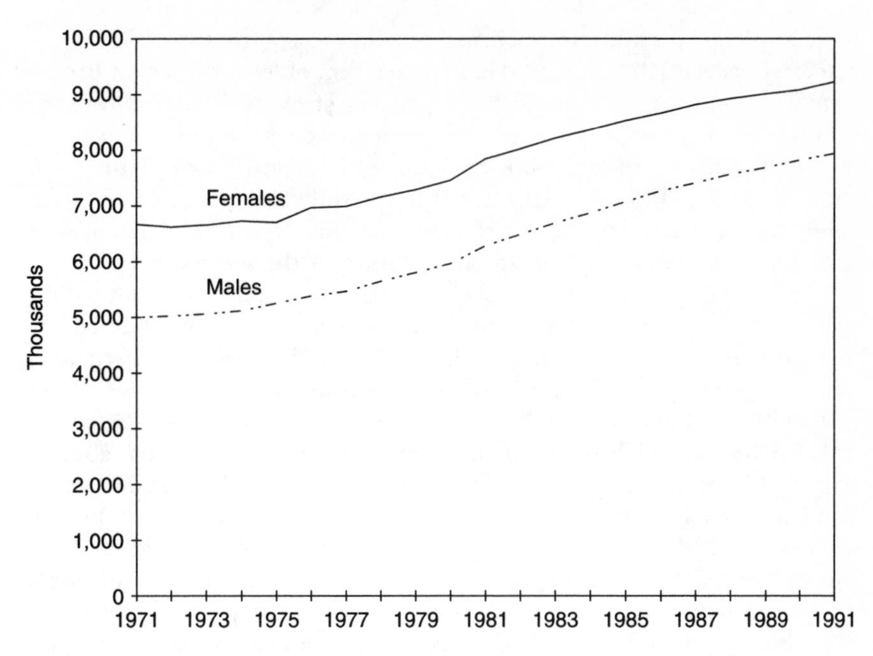


*Figure 5.2* Non-married males and females (aged 16 and over) 1971–91

*Source: Population Trends* 88, table 7

*Note:* Non-married = single, divorced and widowed

approximately 900,000 one-parent families in Britain, but by 1991 the figure had risen to 1.3 million. These families contained about 2.2 million dependent children and represented almost one in five of all families (Haskey 1993a). Ninety-five per cent of these families were headed by women; lone fathers are still quite rare although their numbers have increased marginally in recent years (OPCS 1993d). One-third of lone mothers were single; two-thirds had been widowed, separated or divorced. It is the rise in divorce and separation which has been largely responsible for the increase in one-parent families over the recent past, although since 1988 there has been a significant surge in the number of single mothers (Haskey 1991). The number of families headed by a widowed mother has declined consistently over the past twenty years. There are differences between the types of family headed by single mothers and those headed by separated and divorced mothers. Single mothers tend to be younger and have younger children. The median age of single mothers from 1989–91 was 25 with the modal age of dependent children at 0–4. For divorced mothers the median age was 37 and the modal age of children was 10–15 (Haskey 1993a). Many lone parents face significant hardship in the support of children, partly because of lack of financial resources and partly because of difficulties of finding adequate child-care facilities to allow them to enter the labour market. Between 1986 and 1990 only 40 per cent of lone mothers in Britain were in employment (compared to 56 per cent of married women), and the majority of these were divorced or separated women. Single women not only find it harder to find work but also tend to be in lower-status occupations (Haskey 1993b). It is clear that the different types of one-parent family are generally at different stages in the life-cycle and it may not be appropriate to consider all one-parent families in the same way.

The increase in cohabitation and of single-parent families are both indicative of the general drift away from the traditional nuclear family of legally married parents and children. Another indicator that reflects this trend is the rapid increase in the number of births outside marriage. Between 1981 and 1991 for England and Wales the figures increased from 81,000 to 211,300 births, which in percentage terms represents a shift from approximately 13 per cent to over 30 per cent of total births. The age of mother at birth shows a marked skew towards the younger age-categories. Over 80 per cent of births to women under 20 were outside of marriage in 1991 and 45 per cent to women aged 20–24; whereas only 16 per cent of births for women in the age-range 30–34 were in this category (*Population Trends* 72, table 10). However, the older age-groups showed the largest relative rise in births reflecting the increase in extra-marital births to divorced and separated women and to cohabiting couples (Figure 5.3). Only a quarter of births outside marriage were registered by a sole parent (usually the mother), 54.6 per cent were registered jointly by both parents living at the same address and 19.8 per cent by parents living at different addresses. What these figures indicate is that there is no longer any stigma attached to having children outside marriage and that marriage itself is not necessarily regarded as a prerequisite for family formation. Unmarried teenage mothers are no longer the typical group responsible for births outside marriage, although in general terms mothers of extra-marital births tend to be younger than married mothers. The average age of unmarried mothers at birth in 1991 was 24.84 compared with 28.89 for mothers within marriage.

Although indicative of changing patterns of sexual behaviour, and changing attitudes towards marriage, the statistics for births do not tell the whole story. One in five conceptions, in England and Wales, are legally terminated by abortion. The abortion rate (abortions per 1,000 women aged 14–49) has shown a gradual upward trend over the past twenty years from 8.4 in 1971 to 13.6 in 1990. There are marked differences between the experience of women within and outside of marriage. In 1990 43 per cent of all conceptions occurred outside marriage, as compared with 27 per cent in 1980. Of the 377,000 extra-marital conceptions 36 per cent ended in abortion, whereas only 8 per cent of marital conceptions were terminated. The rate of conception has been rising amongst teenagers, from 59 conceptions per 1,000 women, aged 15–19, in 1980 to 69 per 1,000

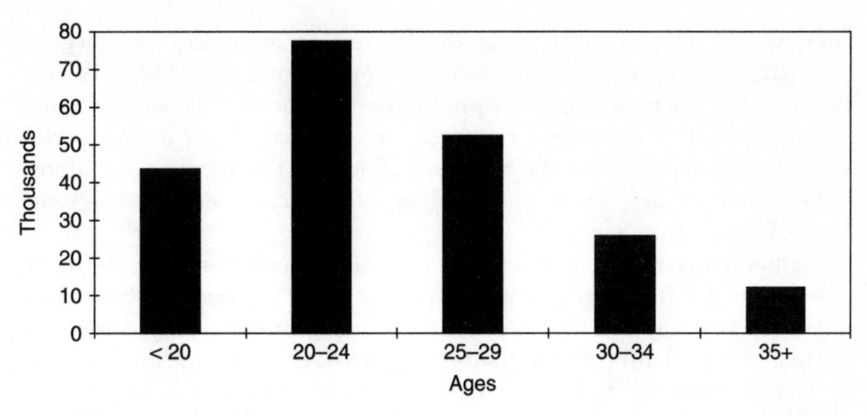


*Figure 5.3* Births outside marriage by age of mother 1991

*Source: Population Trends* 88, table 10

in 1990. Ninety per cent of all teenage conceptions occurred outside marriage, 39 per cent of which ended in abortion. However, the greatest rate of increase in extra-marital conceptions has been experienced by women aged 20–24. In 1990 over half of all conceptions to women in this age-group were outside marriage, again reflecting the shift towards a later age of marriage and the increase in cohabitation. Many more extra-marital conceptions now result in maternity outside marriage as well (OPCS 1993c).

## Births within marriage

Despite the declining significance of marriage as a prerequisite for family formation, the majority of children (70 per cent in 1991) are born within marriage. Alongside the changes that have been taking place in the patterns of cohabitation, conception and child-birth outside of marriage, there have been a number of important developments in the patterns of family formation followed by married parents. In aggregate terms these changes have had a much more important impact on the overall trends in fertility, not only because of the larger number of births involved but also because fertility rates for women within marriage generally remain significantly higher than for single or cohabiting women.

The annual total of births within marriage fell steadily during the 1980s from 553,000 in 1981 to 488,000 in 1991. This was the lowest annual number of births recorded within marriage since 1842 (OPCS 1993c). As indicated, this is partly a function of the rapid growth of extra-marital births, but it also reflects the changing behaviour of women within marriage. The main causes of the decline are a general reduction in average family size and a shift in the timing of child-bearing to the older age-categories; women are bearing less children within marriage and they are having them later in life. This is reflected in the figures for age-specific fertility for successive cohorts of women born since 1945 (Figure 4.6). Over the past twenty years there has been a progressive decline in the size of peaks in fertility for women in their twenties and a significant shift in age-specific fertility towards the later stages in the life-cycle. The number of children born to women over thirty increased by 30 per cent between 1980 and 1990 (Jones 1992).

Figures for average achieved family size also indicate the declining trend, although it is difficult to gain a clear picture of current patterns from these statistics, as many families are still in the process of family formation. Women aged 30 in 1990 had an average of 1.42 children. Twenty years earlier, 30-year-olds had an average of two children and went on to a completed average family size of 2.36 children at age 45. It is possible that a surge of fertility for women in their thirties will result in similar average completed family sizes in the early part of the next century but, despite the shift to later child-bearing, it seems unlikely that this will be the case. The apparent decline in average family size has also been accompanied by an increase in the percentage of women who remain childless. Over 30 per cent of women aged 30 in 1990 were without children. This was twice the equivalent proportion for women aged 30 in 1975 (Jones 1992).

The changing age-distribution of fertility towards the older age-categories has had a significant impact on prevailing trends in fertility. The average age at first birth for women has risen from 25.37 in 1981 to 27.48 in 1991, and a significant number of women are now waiting until over the age of 30 before starting a family. Over 26 per cent of first births were to women aged 30 and above in 1991 compared with 15 per cent in 1981. Delaying the start of family formation has an immediate impact on fertility by limiting the fertile period in women; it also has a longer-term impact by lengthening the gap between generations. In effect the cohort born in the 1960s is spreading out its reproductive capacity over a longer period and thereby reducing the pace of generational replacement. This has had the effect of dissipating, in part, the bulge created by the baby boom of the early 1960s and helps to explain why the increase in the number of women in the age-groups 20–24 and 25–9 has not resulted in a significant increase in the number of births (Figure 5.4).

In association with the shift towards later child-birth has been a reduction in the interval between births. In 1981 the difference between the mean age of mothers at first and second births within marriage was just over two years. By 1991 the gap had narrowed to one and a half years (Figure 5.5). A similar pattern can be identified for the gaps between second, third, and subsequent births. This pattern of compressing the period of family formation is particularly marked for women who delay the start of child-bearing until after the age of 30 (Jones 1992). It is also apparent that, for many women at this stage in life, the decision to begin a family

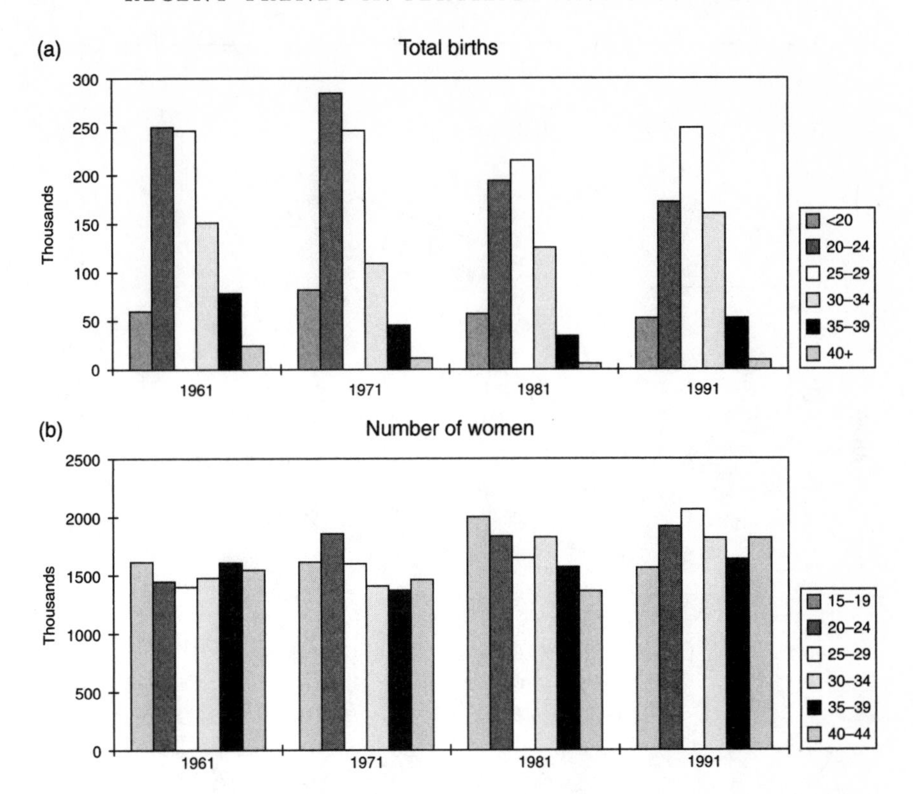


*Figure 5.4* Total births and number of women by age-group (England and Wales) 1961–91

*Sources:* OPCS (1987a), Appendix table 1, *Population Trends* 88, table 9, and OPCS (1993a), table 2

is closely linked to the decision to marry. There has been a sharp rise in the number of births occurring within three years of marriage for women aged over 30 at first birth and the number of births occurring within eight months of marriage (usually conceived outside marriage) has also doubled. For many cohabiting couples, pregnancy has marked an important turning point in a relationship, from informal to more formal patterns of family life.

There are other factors which have also contributed to this pattern of later family formation. For many women family formation may be a feature of a second or later marriage, or remarriage may result in the addition of other children to a reformed family group. The number of births within marriage, to remarried women, increased steadily during the early 1980s but has been decreasing since 1987, partly in line with the declining popularity of remarriage for divorced partners. The mean age at birth, for remarried women, has always been higher than that for married women in

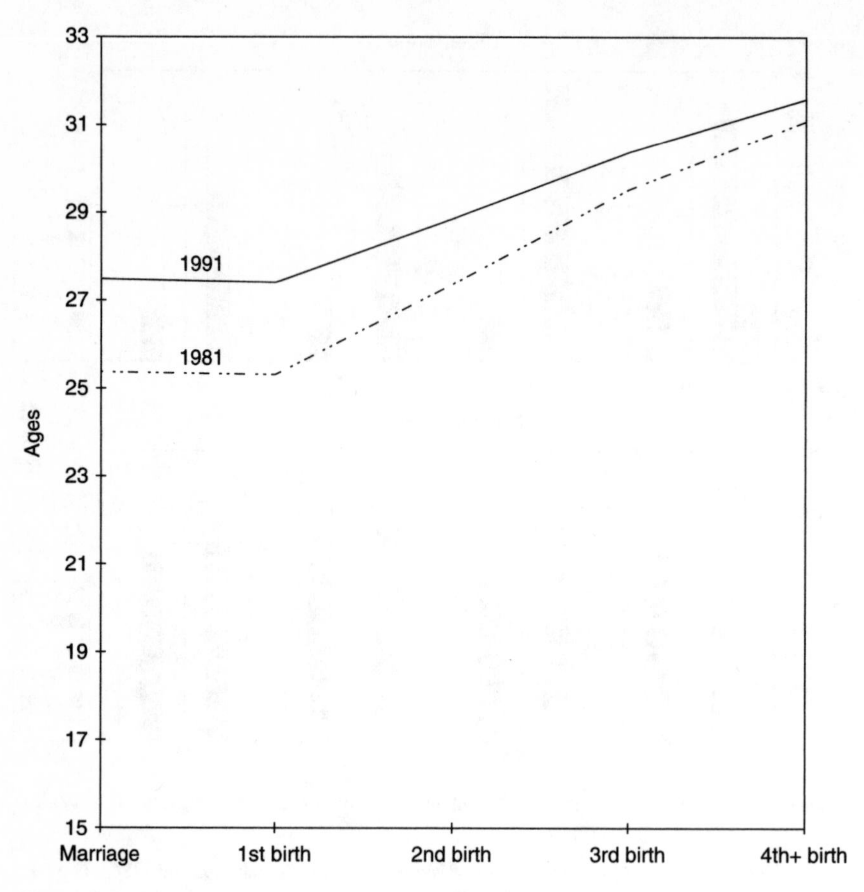


*Figure 5.5* Mean age at marriage and at child-birth 1981 and 1991

*Source: Population Trends* 88, tables 11 and 21

general and over the recent past the rate of increase has also been slightly faster (Figure 5.6). In 1991 69 per cent of births to remarried women were to mothers over the age of 30, compared with 56 per cent in 1981. This group is also responsible for a large share of children born to mothers over 40. Although only 8 per cent of all births within marriage are to remarried women 27 per cent of births to married women over 40 are to remarried women (OPCS 1993c and *Population Trends* 71, table 11).

## Economics of family life

One of the main reasons for the delay to marriage and family formation is the relative expense of children and the loss of potential earnings for women caused by a period absent from employment. Many families have to assess

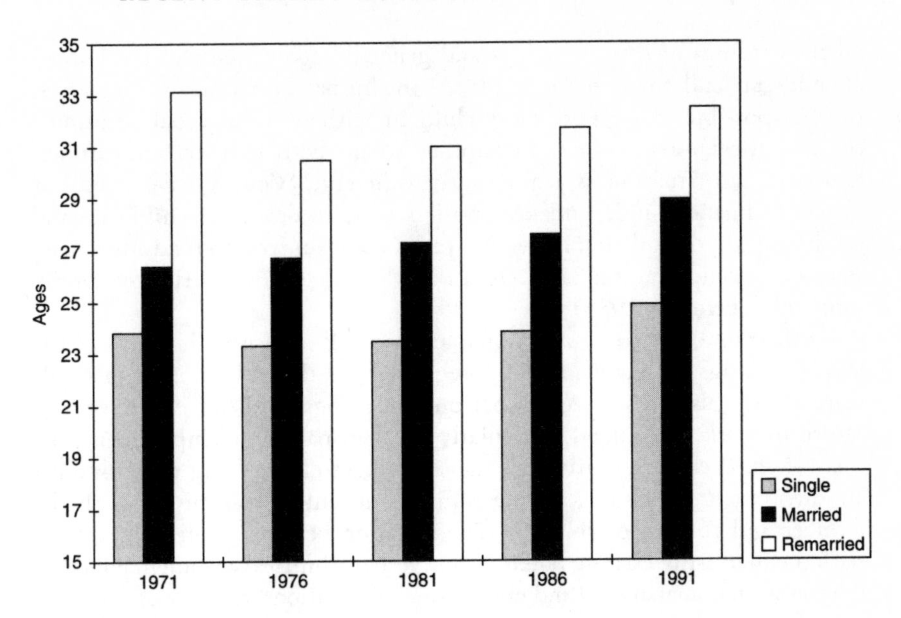


*Figure 5.6* Mean age at child-birth – single, married and remarried women 1971–91

*Source: Population Trends* 88, tables 10 and 11

the costs of supporting children against other expenditures and make rational decisions about when, during the life-cycle, they can afford to start a family and about how many children they can support. State support in the form of child benefit provides only very limited financial assistance to parents and for many represents a fraction of the real costs of raising children. As a consequence, married couples, or cohabiting partners, encounter significant additional costs by taking the decision to have children and for some this may result in marked changes in their material living conditions.

However, for the majority of families it is not the additional expenditure that is the critical factor but the loss of women's income during the period of child-birth and the early stages of family life. There has been a dramatic increase in the participation of women in the labour market over the past twenty years – particularly in part-time employment (see Chapter 7). There has also been a significant improvement in the pay of women relative to that of men, so that in many families the joint income of both partners is either essential to meet basic living costs or, for many young adults, has allowed the attainment of a life-style that they are reluctant to forgo. As John Ermisch has pointed out, this situation has added greatly to the real costs of child-birth. When the income of men is high relative to that of women, there is a tendency for couples to marry at a younger age and have larger families. Women in this situation tend to be dependent on the earnings of their husbands and generally accept most of the burden of child-care and home management. As the husband's real income increases, so it is possible to support more children with only marginal additional costs to the family budget. In families where both partners are earning, however, the situation is significantly different. Women have a greater degree of financial independence and less time to devote to child-care and household responsibilities. Increasing real income for women tends to depress fertility by encouraging later child-birth and a shorter period of family formation (Ermisch 1983).

Furthermore, the costs of children are not only measured in terms of the loss of earnings for women during pregnancy and the early months of child-rearing, but also in the additional outgoings on child-care when women return to work. For many, particularly those in part-time employment, the costs of child-care may outweigh the advantages of returning to work until children reach school-age, adding a further incentive to limiting both the number and spacing of children. The situation becomes more complicated with the increasing earning potential and responsibilities of women. For those in professional, managerial and non-manual occupations, the break from work will often represent a serious interruption to career development and, despite the provisions of maternity leave, may result in the loss of opportunities for advancement. Many women in this situation will wait until their careers are well established before marrying and starting a family. In the analysis of fertility of the over-thirties referred to above (pp. 83–5), nearly half of the births within marriage to women over 30 were in families where the husband was in a professional or managerial occupation. For women over 35 the proportion was even higher at 72 per cent. The results also show that women in this category tend to be more highly qualified (Jones 1992).

## Contemporary patterns of mortality

In comparison to the changing regime of fertility in the 1990s, patterns of mortality are relatively stable and display well-established trends within Britain's population. Expectation of life at birth for England and Wales (see Chapter 3) has risen steadily from 68.1 years in 1961 to 74 years in 1993 for males, and from 74 years to 79.3 years for females. Mortality has become increasingly concentrated into the older age-categories. Eighty-three per cent of deaths in 1994 were accounted for by people over the age of 65. Unlike previous generations, there are now relatively few threats to life in the earlier stages of the life-cycle; Britain has achieved the final stage of the epidemiological transition. The total number of deaths varies from year to year, mainly in relation to the size of different cohorts in the population. Hence the decline in deaths in recent years is as much a reflection of the relatively small cohort born in the 1920s and 1930s (see Chapter 4) as any significant improvement in survival rates.

There are still a number of factors which influence mortality within the short-run. There is a well-established annual cycle to mortality with peaks in the winter months and lower figures during the summer. Very cold spells will result in an increase in deaths, although the effect is generally fairly short-lived. The old are also vulnerable to the effects of infectious disease. In the last major influenza epidemic, over the winter of 1989–90, there was an excess of 30,000 deaths over seasonal averages between 17th November 1989 and 11th January 1990 (Ashley *et al.* 1991). Four-fifths of those who died were over the age of 75. However, in the following three months there was a significant shortfall in deaths. The effect of the epidemic appears to have been to bring deaths forward by a few months, rather than having a significant impact on overall patterns of mortality.

Despite the significant improvements in life-expectancy, the British mortality pattern is not as 'advanced' as some other industrialized nations. The Japanese currently enjoy the longest life-expectancy in the world, with not only a high concentration of death in the older age-categories but also significantly lower mortality levels for both men and women in their fifties and sixties (Ermisch 1990). The reasons for this may in part relate to differences in diet and life-style, rather than state investment in health care, but it illustrates that there is still some scope for improvement in mortality in Britain. In particular there is a need to focus on the causes of death in the 45–74 age-range which in 1994 accounted for 37 per cent of all deaths. Additionally there remains a wide disparity in the mortality statistics of different social groups. Improvements are not shared consistently by rich and poor in Britain.

## Changing trends

The explanation for the general improvement in mortality is the longer-term changes in the causes of death. Over time, the life-threatening infectious diseases have declined in importance as a consequence of improving conditions of life and improvements in health care. The catalogue of diseases which were commonplace half a century ago (diphtheria, tuberculosis, bronchitis, polio) are now far less significant, particularly for the young and middle-aged. The shift in balance has been towards the degenerative diseases of old age. Four principal causes of death account for the majority of adult mortality (Table 5.2): malignant neoplasms (cancer), heart disease, cerebrovascular disease (strokes) and respiratory diseases (mainly pneumonia and bronchitis).

Deaths from cancer have increased significantly and now account for 25 per cent of all deaths. The patterns of cancer mortality differ for males and females. For men the largest category is lung cancer which accounts for approximately 30 per cent of all cases, although the rate has shown a slight decline in recent years as a reflection of changing habits of smoking.

*Table 5.2* Deaths by cause 1993 (selected causes)

| | *Male* | | *Female* | |
|---|---|---|---|---|
| | *Number* | % | *Number* | % |
| All causes | 278,959 | 100 | 298,926 | 100 |
| Circulatory system | 123,713 | 44.3 | 134,439 | 45.0 |
| Ischaemic heart disease | 79,509 | 28.5 | 66,926 | 22.4 |
| Cerebrovascular disease | 22,615 | 8.1 | 38,557 | 12.9 |
| Cancer | 74,490 | 26.7 | 67,965 | 22.7 |
| Colon | 5,229 | 1.9 | 5,763 | 1.9 |
| Pancreas | 2,853 | 1.0 | 3,021 | 1.0 |
| Lung | 21,665 | 7.8 | 10,944 | 3.7 |
| Breast | — | — | 13,026 | 4.4 |
| Prostate | 8,601 | 3.1 | — | — |
| Bladder | 3,215 | 1.2 | 1,552 | 0.5 |
| Leukaemia | 1,915 | 0.7 | 1,640 | 0.6 |
| Respiratory diseases | 41,799 | 15.0 | 49,067 | 16.4 |
| Pneumonia | 20,839 | 7.5 | 33,758 | 11.3 |
| Pulmonary disease | 17,477 | 6.3 | 11,356 | 3.8 |

*Source: Deaths by Cause*, OPCS Series DH2, no. 2 (1994)

Prostate cancers account for a further 11 per cent of deaths and have shown an upward trend over the past twenty years, as have intestinal cancers. Stomach and rectum cancers, on the other hand, have declined (Dunnell 1995). For women the most common site for cancer is the breast, accounting for approximately 1 in 5 of all deaths. There has been a steady rise in breast cancer over the past twenty years, although in the last two or three years there has been a slight decline. Lung cancer is responsible for roughly 10 per cent of deaths for females – not as significant as for males but on an upward trend. Cervical cancer has declined following more extensive screening of women in recent years.

Deaths associated with heart attacks and strokes represent 'failures' of the body system and are the most common cause of death of the elderly. They account for approximately half of all deaths. However, there has been some decrease in mortality from heart disease over the last ten years across all age categories but particularly for adults under 65. Patterns vary with age, sex and life-style. The rate of ischaemic heart disease has typically been higher for men than women and is associated with levels of cholesterol in the bloodstream and smoking. Women experience significantly higher rates of cerebrovascular disease, particularly in the older age-categories. High blood-pressure is particularly associated with the occurrence of strokes. Respiratory diseases reflect environmental circumstances and living conditions. The most significant decline has been in bronchitis and allied conditions which are directly related to smoking and atmospheric pollution. Pneumonia has shown quite marked fluctuations in the death rate for men and women, but with an increasing shift towards the older age categories. It is significant in circumstances where people's natural immune system is ineffective, as for example in cases of transplant surgery or sufferers from AIDS.

Other causes of death have received attention in recent years, not because of their contribution to the total level of mortality, but because they have demonstrated an increased incidence. In the case of infant and perinatal mortality there has been concern about Sudden Infant Death Syndrome (SIDS or 'cot deaths') which increased markedly during the 1970s and 1980s. This increase has been accompanied by a decline in other causes of post-neonatal deaths and may partly reflect changes in registration practice. Nevertheless, the apparent increase caused much alarm at the time and resulted in research and a publicity campaign to encourage parents to alter infant sleeping positions. Since 1987 the figures have reduced significantly although the syndrome is still not adequately explained. The other disease which has received much attention is AIDS (Acquired Immune Deficiency Syndrome). The condition is caused by the Human Immunodeficiency Virus (HIV), which attacks the body's immune system, exposing individuals to infection. Incidence is still relatively limited within Britain: a cumulative total of 15,000 cases of HIV were identified in 1991, almost two-thirds of which were in the four health authorities in London and the South East (*Guardian* 1991b). In the same year the total number of deaths from AIDS was approximately 3,000. Because of its origins in Africa and the nature of its transmission, it has been seen as a disease of minority groups (immigrants, homosexuals and drug addicts). However, recent evidence has suggested that it is spreading to the population as a whole and that official estimates may significantly understate the true picture. Evidence on HIV largely comes from diagnosed sufferers and there are now moves to introduce anonymous testing of blood samples collected for other purposes in order to gain a clearer understanding of the distribution of the virus. Despite the public concern, the recording of deaths from AIDS has not made any significant impact on mortality statistics and has yet to be recognized by the ONS as a separate cause of death.

## Geographical variations in mortality

One of the most enduring features of patterns of mortality in Britain is the marked variation between different parts of the country. At the regional level there is a clear North/South divide, with life-expectancy for males in the North Western Health Authority over three years shorter than those in East Anglia (Charlton 1996). The same picture is presented by regional variations in infant mortality (Figure 5.7). This pattern was identified by the Registrar General in the mid-nineteenth century and is normally taken

*Figure 5.7* Infant mortality by health region 1993

*Source: Key Population and Vital Statistics for Local and Health Authority Areas* (1993), table 4.3

to reflect variations in both environmental and social conditions. A more detailed analysis of local authority districts confirms the basic pattern but identifies the specific importance of high mortality rates in urban areas, including those in the South – notably the inner London boroughs (Champion *et al.* 1996). Such regional variations are perhaps to be expected and in one sense could be simply explained in terms of the differences between urban and rural environments. Cities, particularly northern industrial cities, are associated with atmospheric pollution, inadequate housing and poor social amenities – conditions which encourage diseases (particularly respiratory diseases) and unhealthy life-styles. In contrast the image of the rural South is one of greater prosperity and a generally healthier environment. Such an explanation is, however, too simple. There have been significant improvements in the environmental conditions within urban areas over the past few decades. The Clean Air Act was introduced in the 1950s, since when the worst excesses of domestic coal burning and industrial contamination have been largely eliminated from cities. Traffic fumes may continue to be a potential source of hazard and may account for some of the variation in asthma deaths. But generally a crude association between environment and mortality does not provide a convincing explanation of the geographical variability.

There are particular causes of death which may be associated with specific environmental circumstances. For example, the relatively high incidence of stomach cancer in parts of Wales has been associated with biogeographical factors (Howe 1960). Similarly there has been much speculation about the link between clusters of leukaemia deaths and the location of nuclear installations. But these are individual examples. The underlying causes of the geographical patterns of mortality are more socially determined and relate to people's living conditions and life-chances. The social variation in mortality is as marked as the regional pattern, and much of the geography of mortality can be explained in terms of the distribution of different social groups in Britain. The reworking of mortality statistics by clusters of local authorities with shared characteristics identifies a difference in life-expectancy for men in 1992 from 75.8 for the most prosperous districts to 71.7 for mining and manufacturing districts (Charlton 1996). In addition, detailed analysis of data from the ONS *Longitudinal Study* (see Chapter 2), identifies a polarization of mortality patterns. Taking car ownership and housing tenure as surrogate measures for socio-economic status, Filakti and Fox have shown that despite a general improvement in mortality across all groups during the 1980s, the differentials across groups have increased in relative and absolute terms (Filakti and Fox 1995).

Explaining mortality in terms of social class presents some difficulties in defining which social characteristics impact most significantly on conditions of life. It is not so much the measurable variables of wealth, occupation and housing tenure as the less easily described aspects of 'life-style' and social behaviour. In identifying the reasons for premature deaths from heart disease, the Health Education Authority attributed causes to a complex mix of factors including: smoking, exercise, diet, occupational hazards and the economic and social history of communities.

## Lifestyle and mortality

The links between smoking and health are well publicized and the impact of smoking-related deaths on health service provision is an issue of some controversy. The Health Education Authority refers to the current situation as 'the smoking epidemic' and estimates that smoking is responsible for approximately 110,000 deaths a year. This amounts to roughly 17 per cent of all deaths. Some 285,000 people are admitted to hospital because of smoking and on average occupy 9,500 hospital beds. The total cost of smoking to the National Health Service is £437 million (*Guardian* 1991c).

However, the sale of cigarettes and tobacco brings in approximately £4 billion a year in tax revenue for the Treasury. Because smoking is not itself a disease and is a matter of personal choice, it has become a tolerable cause of death. Rather like road traffic accidents, society has come to accept these causes of death as a part of modern society and outside the scope for government intervention.

Patterns of smoking vary significantly between different social groups. Approximately 15 per cent of professionals smoke in comparison to over 40 per cent of unskilled manual workers (OPCS 1991c, table 4), but all social groups have generally seen a reduction in smoking and more recent reforms have increasingly discouraged the habit in public places. Against the general trend has been an increase in smoking by individual groups, notably young women. Smoking habits among younger teenagers have remained persistent over recent years, with 30 per cent smoking by the age of 16 (Dunnell 1995). The geography of smoking does not map directly with the distribution of deaths from lung cancer and respiratory diseases as other factors are also involved, but the general picture shows an association. The heavy smoking areas are concentrated in the North and West, including Scotland, whereas figures are lowest in the Home Counties, East Anglia and the South West (Coleman and Salt 1992).

The link between exercise and general health are well established, although the contribution to underlying variations in mortality are less clearly identified. Patterns of exercise tend to vary with life-cycle stage rather than social class, with individuals being most active in the teenage and young adult years. However, there is little evidence to suggest that professional sportsmen and women, who achieve peak levels of fitness, end up living longer lives than less energetic members of society. There is a social-class dimension in the sense that exercise activity is more popular with 'middle-class' adults, and the geographical distribution of many sporting facilities shows a predominance of supply in suburban locations and under-provision in poorer inner-city districts. Unfitness and obesity do contribute to premature death and there is an apparent association between weight and social class. A greater proportion of adults in social groups IV and V are overweight in comparison to social group I (Coleman and Salt 1992).

One of the other factors influencing obesity is diet. When Edwina Currie was Minister of State for Health in the mid-1980s, she made a famous comment about the poor quality of northern working-class diets and the over-consumption of 'fish and chips'. Although seen at the time as an unguarded statement, there is some basis in the evidence to suggest that eating habits do vary considerably between regions and between different groups in society. According to the National Food Survey Committee, meat consumption is highest in the North and North West standard regions and in London, and lowest in the South West, Wales and Scotland. Fresh greens, fruit and wholemeal bread all figure more prominently in the 'southern' diet than in the North (Coleman and Salt 1992). Although the patterns are very generalized, and within regions there are differences in dietary patterns in association with life-style and social class, nonetheless there are important local patterns of food consumption which may influence underlying trends in health and mortality. For example, work on fast-food outlets in Liverpool shows a marked concentration in the poorer parts of the city, reflecting an over-reliance on certain types of food (Paulson-Box 1994). A report on the Scottish diet in 1993 also noted that amongst other factors 'the poor diets of low income groups relate to the paucity of suitable food outlets in inner city areas' (Scottish Office 1993: 2).

The relationship between occupation and mortality has become relatively less important, with structural changes to the British economy resulting in relatively less physically stressful and dangerous occupations and relatively more service occupations. In the past there has been a clear association between certain types of work and particular diseases, of which perhaps the best known is the prevalence of pneumoconiosis in coal-mining areas. Many industrial jobs are hazardous, although the contribution of accidental deaths to total levels of mortality has always been relatively small. In more recent years working patterns have come to influence mortality in different ways. The greater pressures of work in the tertiary sector have added to problems of stress and 'burn-out' for executive staff, which impact directly on heart disease, stomach ulcers and other conditions. Not working is also bad for health. Information from the *Longitudinal Study* (Fox *et al.* 1985) identifies that the unemployed experience a significantly worse health record than those in employment.

Past occupational patterns continue to influence mortality in the 1990s because of the long-term effect of adverse conditions. People live with the 'legacy of the past' and the circumstances which prevailed at the time of childhood can have an effect on people's life-time health experience and life-expectancy. The generation in their sixties today are the generation born in the depression years of the 1930s, many of whom suffered from material deprivation – one reason, perhaps, why mortality levels in the age-range 55–75 remain relatively higher than might be expected. The introduction of the Clean Air Act in the 1950s may have removed much of the threat of atmospheric pollution from British cities, but many of those suffering from respiratory difficulties today may have first experienced problems many years ago. The chances of life are determined by the whole life experience and not just the conditions that prevail at the time of old age. The same is true about the associations between place and health. When people move from region to region, they take with them the associations of their place of origin. Migration on retirement from an urban industrial environment to the healthy conditions of a seaside resort does not, of itself, overcome the accumulated experience of previous living conditions. This implies a time-lag between measures taken to improve environmental conditions or changing life-style habits and their impact on prevailing trends in mortality. The real benefits of improvements over the past thirty years may not become apparent until the current adult population reaches old age.

## Morbidity and health

The study of mortality only provides part of the more general picture of people's experience of health. Morbidity deals with the prevalence of disease and sickness and covers a wider range of circumstances than those which cause death. However, the basic geographical patterns and associations with life-style are very similar to those identified for mortality. The census in 1991 collected information for the first time about 'limiting long-term illness', designed to identify the proportion and distribution of people with health problems or handicaps which limit daily activity or work (see Chapter 2). The geography of illness demonstrates a familiar North/South divide, with the highest proportions in the older industrial districts and metropolitan centres, and the lower figures in the Home Counties and other more affluent rural districts (Figure 5.8). The coalfield areas in particular stand out on the map, demonstrating the legacy of respiratory disease left behind by the mining industry. The explanation for the relatively high figures found along the South Coast, and other favoured retirement locations, is the imbalance in age structure. Clearly the elderly suffer disproportionately from problems of ill health.

However, simply to link morbidity to age and occupation only provides a partial explanation of underlying patterns. Much of the variation in illness is associated with living conditions and life-chances. Of fundamental importance is the association between poverty and health. Social-class differences in health mirror those for mortality, with the incidence of disease being markedly higher amongst the least well-off. Moreover, problems of unemployment and deprivation appear to have a marked impact on health. Evidence from the *Longitudinal Study*, which charts people's progress through the life-cycle, indicates a close association between the conditions of life and the conditions of health (Fox *et al.* 1985. The important determining factors include diet, living conditions, access to health care, occupational hazards and 'cultural' factors. This latter category reflects attitudes to health as well as the circumstances which influence the incidence of disease. There is a tendency for poor health to become persistent between generations. Parents who experienced health problems also tend to have children who are less healthy. This indicates the difficulty of studying morbidity in isolation from the social dynamics which influence people's lives, and raises questions about the effectiveness of addressing the symptoms of poor health without addressing the inequality which may be the root cause.

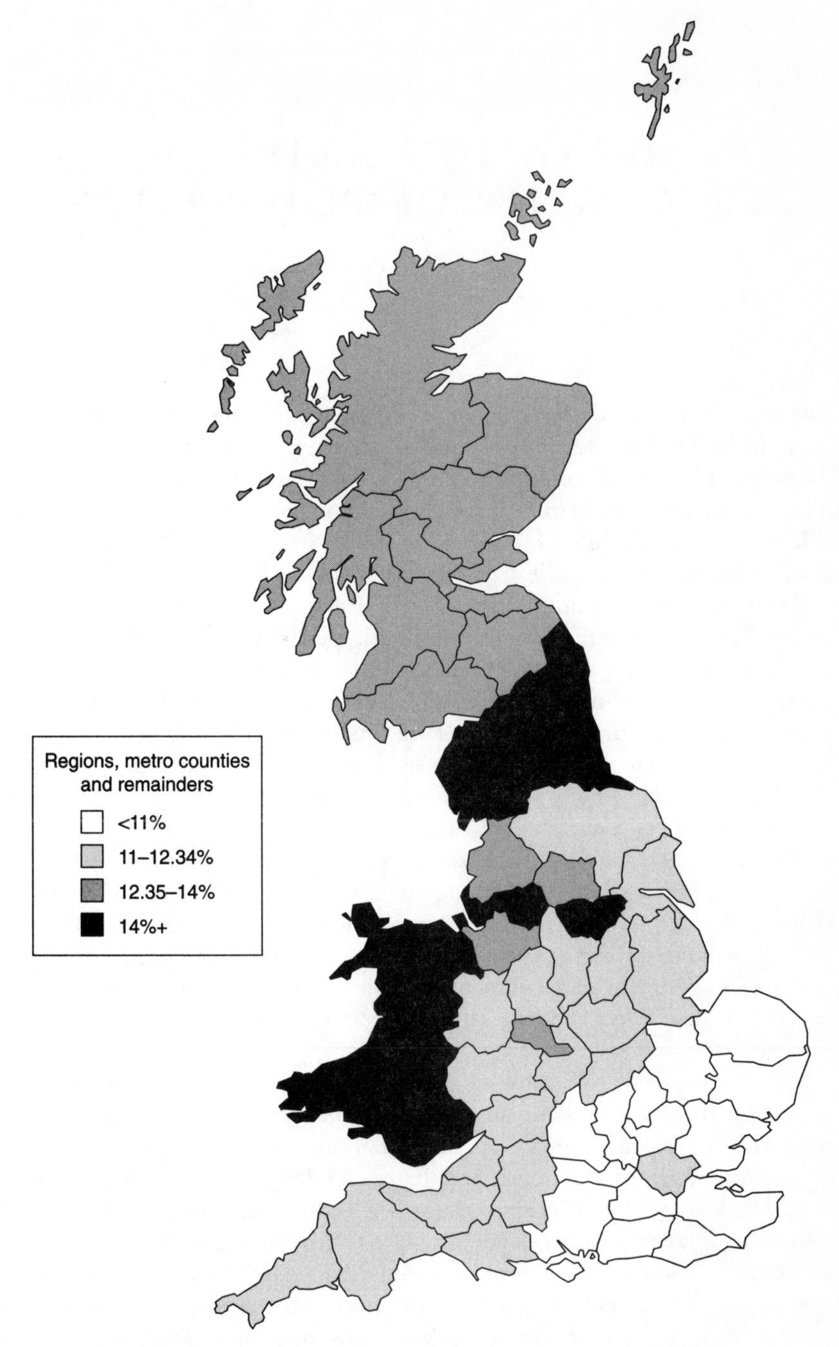


*Figure 5.8* Limiting long-term illness 1991
*Source:* OPCS (1993b)
*Note:* For a definition of limiting long-term illness see p. 8

# 6

# MIGRATION AND THE GEOGRAPHY OF POPULATION

The patterns of population change in any given area are determined by the combined effects of prevailing trends in fertility, mortality and migration. As noted, Britain's population in the 1990s is relatively stable, with little inherent potential for natural growth in the foreseeable future and a probable long-term decline. However, a stable national regime masks the importance of more volatile patterns of change within individual localities. Patterns of fertility and mortality vary from place to place in relation to population structure and local economic performance, but the principal influence on local patterns of change is the movement of population. Some areas of Britain have lost significant numbers over the past twenty years, particularly the inner areas of the major metropolitan centres. Other localities have grown substantially as a result of a sustained influx of migrants. The movement of population not only influences the distribution of total numbers but also impacts on local variations in fertility and mortality. Migration flows tend to be selective in terms of age and social status. The influx to the expanding centres in the South East tends to be characterized by young, skilled or professional adults with the effect of creating a bulge in the age/sex structure and a rise in underlying trends in fertility. The stream of retirement migrants to the English coastal resorts has the reverse effect, adding to the levels of mortality in the receiving communities.

Patterns of population movement occur at different levels and at different scales. In Britain the dominant flows are the general drift of population from the industrial districts of the North and West towards the metropolitan economy in the South and East, and the outflow from the larger towns and metropolitan centres towards suburban and rural districts. This latter stream of movement is commonly referred to as 'counterurbanization' and reflects a sustained pattern of population redistribution that has been effective at varying rates throughout the twentieth century. At a more local level there is a constant ebb and flow of movement over short distances as a consequence of residential relocation. In this chapter the complex patterns of population movement, reflecting both dimensions of time and space, will be considered at a number of different scales: international migration, inter-regional movements, intra-regional shifts and local mobility.

## International migration

In terms of the total impact on population trends, international migration during the 1990s has been of relatively little significance. During the five-year period 1990–94 the UK gained an additional 113,000 as a result of the excess of in-migration over out-migration (*Population Trends* 85, table 17). In the longer term the balance between inflow and outflow has varied, with periods during which the country has experienced a net outflow in numbers (see Chapter 4). However, despite the limited impact in terms of overall numbers, there have been quite significant changes in the geographical patterns of areas of origin and destination of migrants over the recent past. Figure 6.1 illustrates the extent of these changes over the period 1981–94.

In terms of the inflow of migrants, the most significant change has been a decline in the numbers from the New Commonwealth (and Pakistan). This reflects the increasingly stringent controls on immigration in recent years. In contrast there has been an increase in the proportion of migrants from the European Union, at least in part a response to the removal of restrictions on labour migration. There has also been a growth in the numbers from the Old Commonwealth (Australia, New Zealand, Canada and South Africa) and the US. In total the numbers are relatively small but the trend indicates the changing relations with countries outside the European Union. The so-called 'brain drain' across the North Atlantic is not as apparent in the 1990s as perhaps it had been in the 1950s and 1960s.

The pattern of migrant destinations indicates similar longer-term trends in population movement. The most significant decline has been in the numbers emigrating to the Old Commonwealth – over 100,000 in 1981 and down to approximately 40,000 in 1994 (*Population Trends* 85, table 19). The impact of economic recession in the early 1980s and 1990s has effectively closed the opportunities for labour migration from the UK, except in a limited range of skilled and professional occupations. The biggest growth in numbers has again been within the European Union, reflecting both the easier movement of labour and increasing trends for retirement migration to France and Spain. During the 1980s migration to the Middle East was also significant as a consequence of rapid economic expansion, although numbers have declined in recent years.

Although the numbers involved in international migration are small, the patterns of movement do have an impact on individual localities. The sustained migration from the New Commonwealth since the 1950s has led to the buildup of significant ethnic populations within British cities.

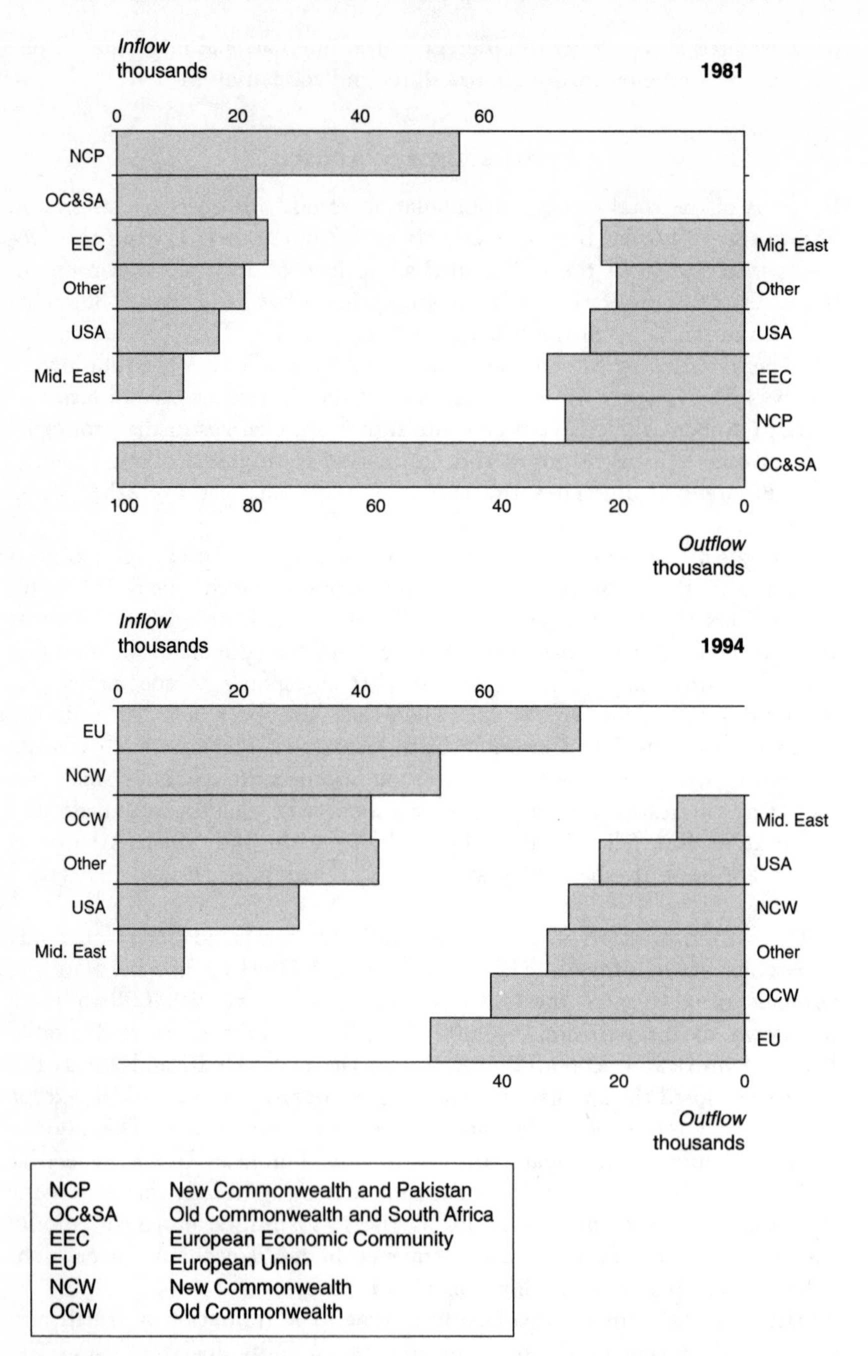


*Figure 6.1* Origins and destinations of migrants 1981 and 1994

*Source: Population Trends* 88, table 19

Different communities are associated with particular areas of origin, and with particular occupational opportunities – for example, the concentration of migrants from the Indian sub-continent in West Yorkshire and East Lancashire in association with the textile industries. Ethnic populations are disproportionately concentrated in urban centres. Thirty-one per cent of Britain's white population live in Greater London and the metropolitan counties. The comparable figure for ethnic populations is 71 per cent (HMSO 1991: 11). A large proportion of new migrants arrive in the South East of England – particularly in London (Coombes and Charlton 1992). Over time they disperse through the urban hierarchy in response to employment opportunities and kinship and community links. Whereas individual metropolitan counties may be typified by specific ethnic populations, the capital has representative groups of all migrant populations. Variations in the numbers of migrants arriving in the UK can have an impact on the population of London. During the mid-1980s, when London's population grew marginally, one of the factors identified was the influx of new migrants (Champion and Congdon 1988).

## Inter-regional migration

Since the Second World War patterns of population change in Britain have been strongly influenced by trends in migration between the regions (Figure 6.2). In particular there has been a relative shift of numbers away from the former manufacturing areas in the North and West of Britain towards the relatively prosperous metropolitan economy of the South East. The standard planning region of the South East (including Greater London) now contains nearly one-third of Britain's population and continues to attract in-migrants at the rate of 250,000 a year (Table 6.1). This inflow is matched by a counter-current of migration from the metropolitan core towards the periphery – a process of displacement that has resulted in the outward expansion of the geographical area of the metropolitan economy into East Anglia and parts of the South West. The balance between these two major currents of population has varied over time in relation to the relative performance of different regional economies. For example, during the period of relative economic expansion in the South East in the mid-1980s there was a rise in the numbers attracted to the centre. However, in the recession years which followed in the late 1980s and early 1990s, the trend was reversed and the region experienced a net loss of population.

The migration flows are very different in terms of the people involved. In-migration is dominated by young adults and, according to Fielding (1989), contains a disproportionate share of Britain's skilled and young professionals. Out-migration tends to be characterized by 'middle-aged professionals and managers' with their families, and by retirement migration (Fielding 1989). The South East is seen as an 'escalator' region, providing

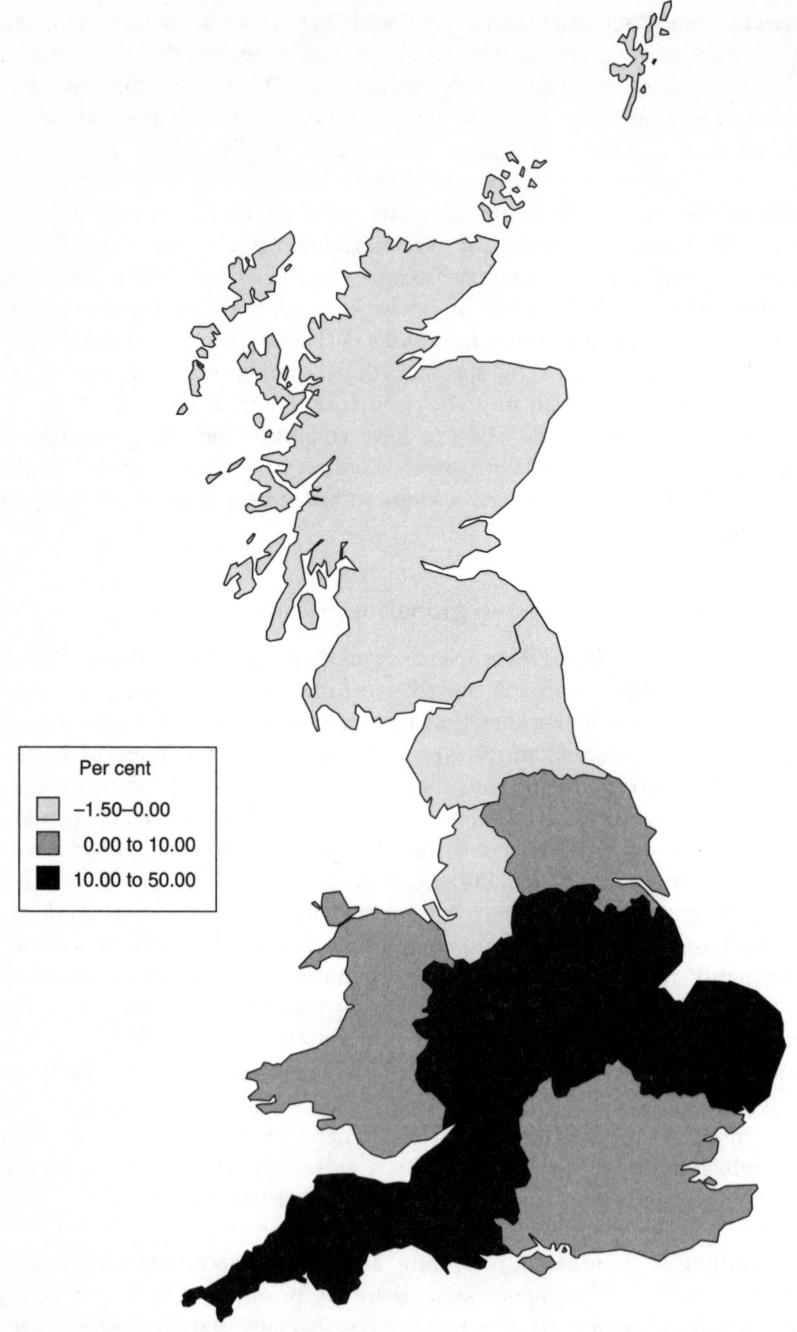


*Figure 6.2* Population change 1961–91

*Source: Regional Trends* (1995), table 3.1

*Table 6.1* Population shift by region 1991

| | *Inflow* | *Outflow* (*thousands*) | *Balance* |
|---|---|---|---|
| England | 95.8 | 112.2 | –16.4 |
| Wales | 51.5 | 47.4 | +4.0 |
| Scotland | 55.8 | 46.7 | +9.2 |
| North | 49.8 | 49.5 | +0.3 |
| Yorks and Humberside | 85.0 | 85.4 | –0.4 |
| East Midlands | 89.6 | 81.4 | +8.1 |
| East Anglia | 58.1 | 47.7 | +19.1 |
| South East | 223.1 | 264.7 | –41.7 |
| Greater London | 148.8 | 202.1 | –53.3 |
| Remainder | 249.9 | 238.2 | +11.7 |
| South West | 120.7 | 98.9 | +21.8 |
| West Midlands | 82.7 | 87.9 | –5.2 |
| North West | 90.1 | 99.9 | –9.8 |

*Source: Population Trends* 88, table 20

opportunities for advancement for young adults, but with many cashing in on the possibilities of promotion, business ownership and self-employment or simply a better quality of life, later in the life-cycle, by moving away from the capital.

Although the ebb and flow of population between the 'southern' economy and the more peripheral districts of Britain may represent the principal undercurrent of inter-regional population shift, there are also significant cross-currents between other regions. Over recent years there has been a significant growth in the populations of the South West and East Anglia. The South West has expanded from a population of 3.7 million in 1961 to 4.8 million in 1994. Only part of this expansion can be accounted for by 'overspill' from the South East; it also reflects a growing popularity of the region for retirement migration and the relative success of local economies in places such as Bournemouth and Bristol. The relative growth in the population of East Anglia is based on a smaller total population (2.1 million in 1994) and in part represents displacement from the South East. Some of this is 'direct' overspill in the form of new and expanded towns during the 1970s, but it also indicates the growth of new centres outside Greater London. In particular Cambridgeshire has grown consistently in response to newer economic initiatives and the expansion of East Anglian ports – with increasing European trade – has helped to create a new centre of economic gravity in East Anglia.

Against a backdrop of a stable national population the regional shifts in population in effect represent a 'zero sum game'. The gains in one area are matched by losses elsewhere. The growth in the population of the Southern regions has had the effect of creaming-off the natural increase in other parts of the country. The main areas of loss have been the larger metropolitan areas of the North and West and the former industrial districts, although the transfer in population has not necessarily resulted from a direct flow from areas of relative underperformance to areas of economic expansion. The causes of movement are far more complex in terms of their impact on individuals. The decisions to move may be influenced by a large number of personal and specific reasons. Aggregate trends, however, may be accounted for by a number of principal factors.

Variations in regional economic performance set the scene for underlying trends. As noted by Ravenstein in the nineteenth century, the chief stimulus for migration is invariably economic 'betterment' – people move in response to opportunities (Ravenstein 1885). The restructuring of Britain's economy over the past thirty years has resulted in a decline in employment opportunities in the former industrial regions and an expansion in 'new' jobs in the service and tertiary sectors in the southern economy. This macro-level shift in the geographical focus of Britain's economy has encouraged population movement – although the process is not immediate. The effects of varying regional activity result in longer-term trends in the shift in numbers.

The actual level of movement at any given point in time relates to the state of both national and local economies. Generally people are more mobile during periods of economic expansion than recession. Economic opportunity has a greater influence on migration than does the effect of de-industrialization and employment loss. The impact of the recession in the early 1980s in Britain led some commentators to suggest that those disadvantaged might overcome hardship by looking for work elsewhere. But in reality the unemployed are amongst the least mobile groups in society. They are disadvantaged by a lack of appropriate skills and by the fact that their experience is rarely of value in new sectors of the economy. In addition, the problems associated with finding affordable accommodation, and the ties of family and community, all mitigate against a free flow of labour within the national economy. It is those with the more advanced education and training who tend to move away, further depleting depressed regions and hampering prospects for economic redevelopment.

Other factors which influence patterns of inter-regional movement include population structure. Life-cycle stage is significant in determining who will move and when. A significant amount of migration occurs at 'transitional' phases in people's lives – particularly at the transition between education and employment and the transition between employment and retirement. Young adults are the most mobile group in the population, not only because of the diverse opportunities available for entry into the labour market, but also because of their general lack of commitments of home ownership and family rearing. As a consequence growth regions tend to attract a disproportionate share of young adults, and localities which are losing numbers become unbalanced in their structure. The destinations for retirement migrants similarly reflect the effects of age-specific migration, with an increasing burden placed on the health and social services for the care of an ageing population.

## Intra-regional mobility

Patterns of population movement between the regions may reflect the longer-term process of population redistribution in Britain. However, the majority of moves are over relatively short distances. Over 70 per cent of moves in Britain are over distances of 10 km or less – most of which would come under the heading of residential relocation rather than migration (Owen and Green 1991). People move house for a variety of reasons, depending on individual circumstances, but it is possible to identify aggregate patterns of movement from the overall volatility of local moves. One of the well-established trends is the process of 'suburbanization', brought about by a constant drift of population from the central areas of towns and cities towards newer housing developments on the periphery. This is not a new process and can be identified in Britain from the peak period of urban growth in the nineteenth century.

In the ecological model of urban growth developed in the 1920s by Park, Burgess and McKenzie, the shift of population to the fringes is seen as a consequence of social mobility and rising wealth. Individuals and families attempt to find the optimum location in relation to their financial circumstances, resulting in a social gradient from the run-down inner-city locations to the wealthy, salubrious suburbs. In the model this creates a constant pressure for the outward expansion of urban areas as the conditions of individuals improve and the position of the less well-off members of urban society is replaced by a steady inflow of migrants to the inner-city districts (Park *et al.* 1925). The ecological model describes a situation in North American cities between the wars, but some of the basic principles apply to contemporary towns and cities. There is a general association between wealth and location, with much suburban development typified by 'middle-class' property and perceived as 'desirable neighbourhoods'.

Real income is not the only determining factor behind population movement within cities. Other important considerations include life-cycle stage and family arrangements. Suburban locations are favoured by younger families because of the facilities for children and, in some areas, because of considerations about school provision. Single adults are far less influenced by the supposed attractions of suburban life and may choose to locate in the central areas of towns, particularly in newly gentrified developments which may be close to their place of employment and which allow access to the attractions of the central districts. Retirement is another important stage in the life-cycle, which can prompt relocation within urban areas, with many people moving to locations away from former places of employment to pleasanter surroundings in the suburbs.

Social forces may have a continuing influence over local patterns of mobility, but they are not the sole factors responsible for the distribution of population in urban areas. Since the Second World War both central and local government have intervened in the development of urban areas, both through the application of planning restrictions and through housing policy. During the 1950s and 1960s there were large-scale plans for the restructuring of towns and cities, which resulted in the relocation of population from inner-city areas to newer out-of-town housing estates. This in part was a reaction to the need to improve significantly the basic living conditions and remove the worst of the poor-quality housing in central districts, but it was also an attempt at social engineering, to provide a better living environment in suburban districts for inner-city families. The failure of many of these estates was partly due to faults in the design of the new tower-block properties, but also to a lack of provision for social and community development. An under-provision of facilities resulted in many new residents retaining links with former communities in the city centre. In more recent years local authorities have been reversing the trend of encouraging relocation to the suburbs by refocusing on the development of inner-city areas.

During the 1980s and 1990s the process of suburbanization has been overtaken by 'counterurbanization', which represents a significant shift away from urban forms of living. The process of counterurbanization involves not only the movement of people away from cities but also a decentralization of urban functions; it is accompanied by the rising importance of small towns and rural locations in distributed patterns of industrial activity. In one sense this may be seen as an inevitable progression from suburban growth – with people moving further and further away from the urban core, but there are other factors which help to explain the trend (Fielding 1989).

In terms of physical development, the process of suburbanization had become increasingly constrained by the imposition of planning controls. 'Green belt' policies adopted by local authorities were designed to stop the remorseless spread of suburbia and preserve a divide between town and country. In effect they have acted as a girdle around the major conurbations and have created 'fire breaks' in the outward expansion of the built-up area. In order to maintain a supply of new housing, particularly during the boom years for property sales during the 1980s, developers have been forced to choose locations beyond the fringes of the green belts and thereby encourage the expansion of smaller settlements.

A related factor was the rapid increase in house prices during the later 1980s, particularly in the South East (see pp. 115–18). The value of property varies significantly with location and the most expensive houses are generally found in suburban areas. Those entering the housing market during the boom years were effectively prevented from buying property in the more expensive suburban areas, and as a consequence were forced to consider housing at a greater distance from the major conurbations. Linked to this have been improvements in communications which have made possible longer-distance patterns of commuting – for example, the electrification of the East Coast main line, which has brought the East Midlands within the commuting zone of London.

These factors may help to explain the discontinuity in suburban expansion, but it is also apparent that for many people the move to a more rural location represents a reaction against the conditions of urban life in the 1990s. There has been a growing disenchantment with cities as places to live because of concerns about crime, congestion, pollution and social malaise. Locations in villages and small towns are perceived as offering a better quality of life. Increasingly people are able to have the benefits of both worlds, by enjoying the attractions of life in the country whilst retaining access to the employment opportunities, services and functions of the larger urban centres.

## Population flows within the South East: a case study

Patterns of population change within the South East region show significant variations between places and highlight the importance of shorter-distance movements as opposed to inter-regional shifts. The pattern has changed significantly over the last three decades, as shown in Figure 6.3. During the 1960s the classic pattern of population decline in the urban core and growth in the outer-suburban areas was apparent. The areas of expansion stretch into East Anglia, mid-Kent, parts of the South Coast, the South West and an extensive segment of the Home Counties to the west and north of London. The rapid rates of change in these areas are partly explained by the national trend in population increase during the decade. Between 1961 and 1971 the UK population increased at the rate of 0.6 per cent per annum (*Regional Trends* 1991, table 4.1). In addition it was a time when a large proportion of London's population was displaced by the clearance of sub-standard housing from the inner areas. Many of those involved were relocated to the new and expanded towns within the region, which achieved their highest rates of development during this decade. Economic prosperity in the South East was also drawing a large number of migrants from other regions, many of whom found employment in the developing light-industrial and service occupations of the more prosperous areas surrounding the capital. Slower rates of growth (less than 10 per cent from 1961 to 1971) were experienced on the fringes of Greater London. The inner-suburban ring was already becoming overcrowded and

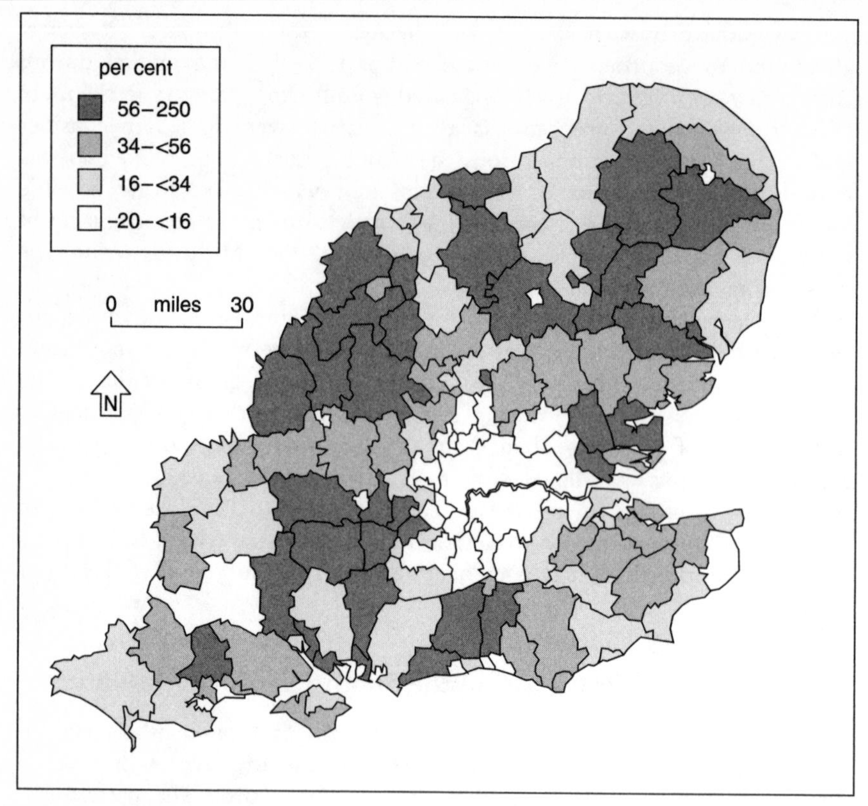


*Figure 6.3* Population change in the South East 1961–91

*Source:* OPCS (1991a)

new development was constrained by 'green belt' policy (Champion 1989). Also the margins of the region (North Norfolk, East Kent, Dorset) were relatively remote from the centre and lost significant numbers through out-migration.

The 1970s still revealed this basic pattern, but with significantly lower levels of growth. The impact of economic recession in the early part of the decade, coupled with the sharp downturn in fertility for the country as a whole, took much of the heat out of the rapid rates of population increase within the region. Growth rates in excess of 20 per cent were confined to individual locations. Places such as Thetford, Cambridge, Peterborough, Northampton, Milton Keynes, Poole, Basingstoke, Bracknell and Camberley continued to do reasonably well, partly as a consequence of the decentralization of economic activity from London, and partly in response to investment in newer, high-tech industries. The area to the west of London, along the axis of the M4 motorway, and Cambridge were particular examples of these newer centres of industrial production. London itself continued to lose population throughout the decade, and the zone of population loss also extended into the districts of the inner-suburban ring (including such places as Reigate, Epsom, Leatherhead, Dartford, Brentwood, Welwyn Garden City, Egham and Slough).

During the 1980s there was a further slowing down in the rates of population growth and a relative shift in population towards the outer edges of the region. Although the middle years of the decade are generally seen as a period of prosperity when the southern economy expanded rapidly, this did not have a direct impact on population trends within the region. This was partly because of continuing low birth rates and partly because of disincentives towards inward population movement (notably a general shortage of accommodation and rapid house-price inflation). Growth points within the region were more widely spread. Only Milton Keynes, which continued its planned expansion during the 1980s, achieved a growth rate in excess of 20 per cent. Other areas which experienced rates of increase between 10 and 20 per cent include the M4 corridor, Cambridgeshire, Suffolk, North Norfolk and Dorset. Significantly, many of these areas were relatively remote from London and their expansion is explained in part by outward migration. The core again experienced a decline in numbers, over the decade as a whole, and the area of loss expanded further into the Home Counties, particularly along the Thames Valley and to the north of London.

Taking the thirty-year period as a whole, the pattern of change in the greater South East appears to reflect the characteristics of metropolitan growth experienced more generally in Western Europe and North America. Suburban growth in the immediate vicinity of London was replaced by 'spillover' as the commuting field of the capital was expanded by improvements in communications. The process was accompanied by the decentralization of economic activity and population from the core to the ring of the metropolitan region. Decline at the centre was matched by growth on the edges as the capital spread its influence over a wider and wider area. This pattern has been identified as an important stage in the general process of urban development, brought about by the increasing congestion and the diseconomies of scale experienced in the core and inner-suburban ring (Van den Berg *et al.* 1982). It also provides support for the counterurbanization thesis. As clarified by Champion, the process of counterurbanization involves a 'shift towards a less concentrated pattern of population distribution, during which a negative relationship between the population size of a place and its rate of net migration change will generally prevail' (Champion 1989: 84). Clearly, during this period there was a significant redistribution of population towards the smaller towns and rural areas at some distance from the capital, beyond the inner-suburban ring. The centrifugal force of decentralization left behind a declining core area which spread further and further into the Home Counties.

It may not, however, provide evidence of the rejection of urban ways of life implied by the strictest definition of counterurbanization. The outward expansion of the metropolitan region resulted in the wider distribution of urbanization across the region. Many of the new growth centres in the outer ring still relied on direct linkages with London and developed a complex web of interaction with other centres within the wider urban network. The process of counterurbanization does not necessarily imply a significant alteration in the operation of the functional region, but rather a shift in emphasis from concentration at the core to more distributed patterns of economic activity (Congdon and Champion 1989). Evidence of counterurban development can also be found beyond the fringes of the region, with population growth in central Wales and the South West of England, although a significant proportion of this growth represented the outward migration of people in the later stages of the life-cycle (Warnes 1992). In addition, many of those who chose to reject urban life and take up residence in remote rural areas still remained in contact with the core via long-distance commuting.

The map of net migration by county for 1988 (Figure 6.4) indicates that population movement, rather than natural change, has been the dominant factor in achieving the redistribution of population within the Greater South East. Variations in natural change between different districts may have been a contributory factor, but generally speaking these tend to reflect underlying differences in the age-profile of places rather than any significant spatial variations in the prevailing trends of fertility and mortality. The overall trend in intra-regional migration is made up of many different patterns of movement for different groups in the population. Net migration patterns for different age-groups show contrasting directions of movement for young adults and the elderly. For young adults the main destinations include Cambridgeshire, Buckinghamshire, Oxfordshire and East Sussex – with Greater London receiving a positive balance of 6.2 per 1,000 (*Key Population and Vital Statistics for Local and Health Authority Areas* 1990, table 5.1). The trend in movement tends to be from the periphery to the core and to the expanding localities around the capital (with the exception of Hertfordshire which consistently lost population in all age-groups). This partly reflects the opportunity in the capital for new entrants to the labour market and for those entering higher education. It is also an indication that young adults without children are more mobile and are prepared to move over longer distances (Coombes and Charlton 1992).

The patterns for the older age-categories present a mirror image of young adults – loss at the centre and gain on the edges. This is particularly true of the post-retirement groups, for whom destinations such as Norfolk, Dorset and the Isle of Wight appear to be particularly popular. The picture confirms the influence of life-cycle stage on the circulatory nature of migration within the region. As Fielding noted, for patterns of

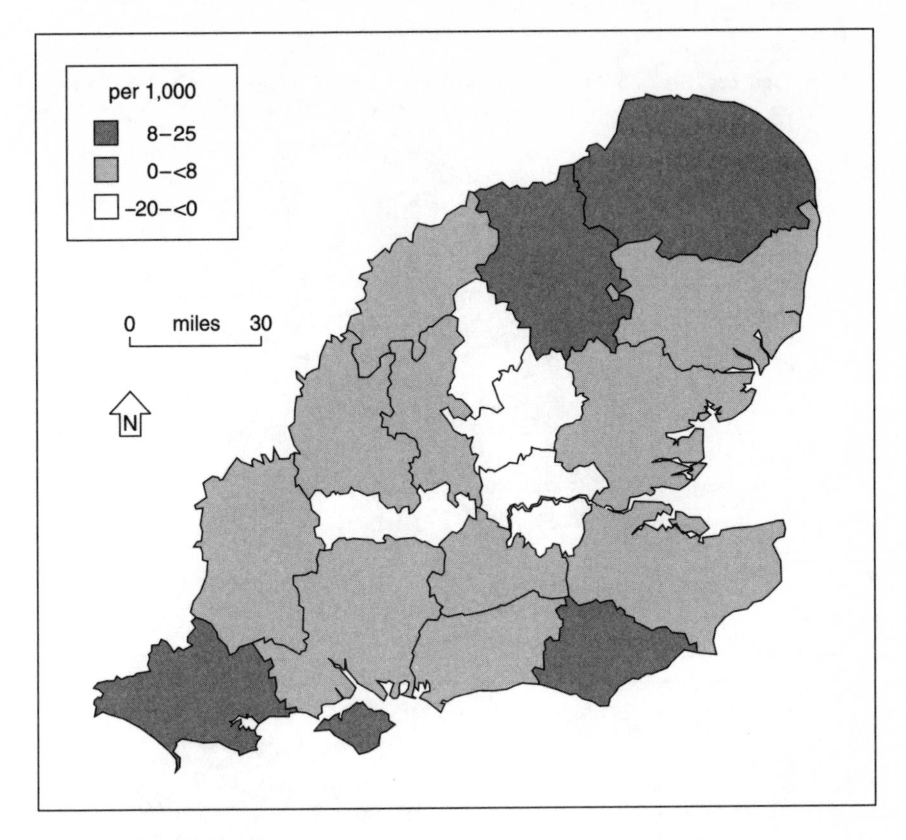


*Figure 6.4* Net migration by county in the South East 1988

*Source: Key Population and Vital Statistics for Local and Health Authority Areas* (1990), table 5.1

inter-regional migration there is an 'escalator effect' in operation – a constant influx of skilled and young professional migrants towards the centre, and an outflow of middle-aged professionals plus managers and their families, together with the retired, towards the edges (Fielding 1989).

The consequence of these differential patterns of movement can be seen in the uneven distribution of different age-groups across the region (Figure 6.5). Young adults predominate in the expanding centres of the outer-suburban ring – particularly to the north and west of London. They are also over-represented in the capital and in other parts of the commuting zone. The elderly by contrast are found either along the south and east coasts, in the South West and East Anglia and in the inner-suburban ring.

This overall pattern tends to confirm the generally accepted view about the relationships between migration and urbanization. Young adults are attracted towards the centres of expansion and opportunity and, at the same

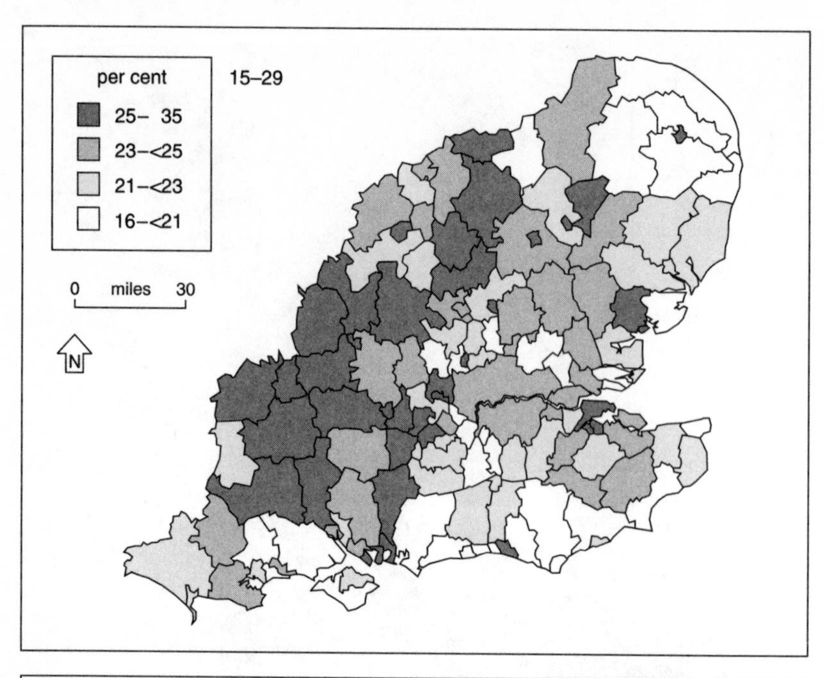


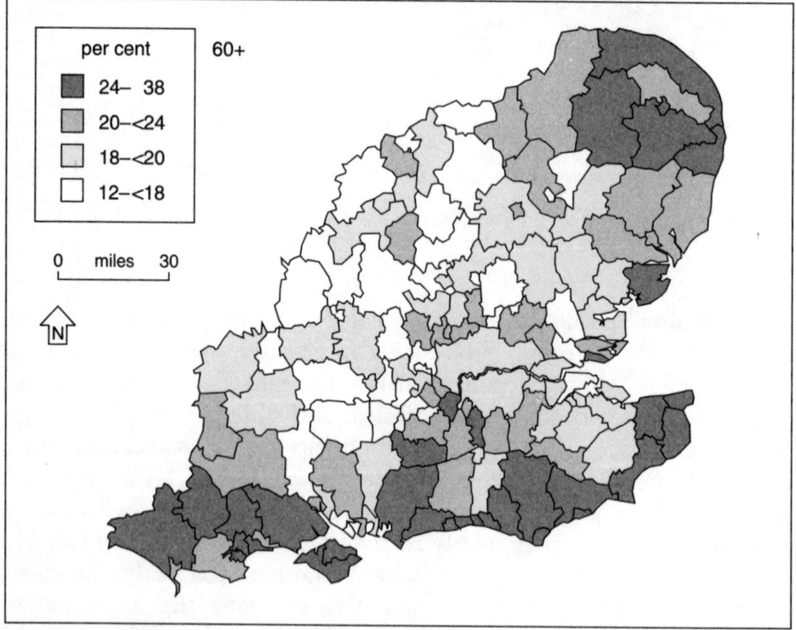


*Figure 6.5* Distribution of age-groups in the South East 1988

*Source: Key Population and Vital Statistics for Local and Health Authority Areas* (1990), table 2.1

time, the forces of social mobility tend to displace people, later in the life-cycle, further and further away from the urban core (Woods 1979).

## Counterurbanization

However, such a generalized explanation does not adequately explain the complexity of the actual patterns of population movement at the scale of local areas within the South East. The pace and direction of population movement have been complicated by the changing geographical structure of the region. The forces of counterurbanization have altered the basic parameters of the process of centralized urban growth and diverted some of the more traditional migration flows in new directions. Increasingly, opportunities for employment have been located in the newer centres of development, away from the capital, including places like Bournemouth which have traditionally attracted large numbers of retirement migrants.

Aggregate patterns of migration are made up of a host of decisions by individuals about when and where to move. These decisions are often influenced by specific sets of circumstances, relevant to particular places, and particular points in time. Within the South East, individual areas present both incentives and disincentives to movement for different groups of people and the resulting pattern of migration reflects the way in which these factors interact with one another. The basic determinants of people's life-experiences include such fundamental things as availability of employment and other opportunities, affordable housing, accessibility to place of work, desirability of residential areas, social networks, basic security, etc., all of which are traded off against one another in the search for the best locations.

The unequal distribution of these factors creates an undulating terrain of opportunity which influences the direction of population movement. At the same time the movement itself is a regulatory device influencing the relative advantages of one locality over another. This has been apparent in the operation of the housing market. Areas which have received significant levels of in-migration have also experienced increased demand for housing which, because of limitations on the development of new property, has generally resulted in rapid house-price inflation. This in turn has altered the attractiveness of the area to potential migrants. Those who can no longer afford the property are forced to look elsewhere, whereas for others rising prices are associated with changing social tone and an enhancement in the desirability of the area. Hence gradually, over time, the character of a locality can be altered as different groups are attracted to move in. Similarly the growth of new centres of production in the outer-suburban districts has resulted in an influx of skilled labour from other areas. This has created shortages in the less rapidly developing districts (such as Thanet and the Medway towns), further limiting possibilities for sustaining redevelopment and exaggerating the differences in economic potential between localities.

The recognition of the fact that intra-regional patterns of migration present a more variegated and fluid picture of population movement at the local scale than normally associated with general models of urbanization and counterurbanization, opens up a number of possibilities for more detailed investigation. The limitations of the available data prevent in-depth analysis of some of the points raised, but it is possible to speculate about the possible impact of individual factors on the changing population structure of the region during the 1980s. In particular four topics have been selected for discussion: changes in the structure and function of London; the regional housing market (particularly the effects of house-price inflation), changing patterns of transport and communications; and, finally, the importance of newer centres of economic development outside the capital.

## Changing trends in London's population

Demographic trends in Greater London have shown marked variations over the past ten years. In an article in *Population Trends*, Champion and Congdon (1988) identified the fact that there had been a reversal in the established trend of population loss from the capital. Between 1983 and 1986 Greater London's population increased by approximately 20,000 – not much, admittedly, but enough for them to suggest that a 'significant watershed' had been reached and to call into question the validity of the counterurbanization thesis. The detailed investigation of the patterns of change during the mid-1980s revealed that this return to metropolitan expansion was associated with the general increase in population in the South East and with increased levels of international migration. Within London much of the new growth was focused on inner areas, most notably Tower Hamlets which contains the Docklands development. This repopulation of the core contradicted the idea that cities were dying from the centre outwards, and instead suggested a more cyclical pattern of urban development with periods of urban decline followed by periods of secondary urban growth – setting up ripples of expansion and contraction that spread out across the wider metropolitan region.

Other factors which may help to explain this reversal in trends relate to the recovery of the capital's economy in the mid-1980s, and the in-migration of young professionals. Also changes in central government policy towards local authority housing provision and overspill programmes reduced the rate of displacement of population from inner areas. In addition, Champion and Congdon (1988) suggested that the rest of the South East was becoming too full. The constrictions of the capital's green belt coupled with the reduction in new house building and house-price inflation, meant there were less opportunities for the upwardly mobile to move out.

Reviewing the situation several years on, this experience of the mid-1980s may not signal a significant reversal in the trends of urban development.

Since 1987 London's population has again gone into decline and, taking the decade as a whole, the relatively small gains of the period 1983–7 are more than offset by the losses in other years (total loss 1981–91 = 320,000). It is difficult not to relate this to the effects of economic performance over the decade, with the recession of 1990–94 having an impact on employment opportunities in the capital and contributing towards a marked fall-off in in-migration from other regions of the UK (*Population Trends* 72, table 20). However, even if the period of growth is not seen to be a significant turning-point in the process of urban development, it nonetheless illustrates the point that growth and decline are essentially determined by economic opportunity – not by natural ecological processes of urban decline, or counterurbanization. The process of repopulation was not restricted to London but was also identified in other European and North American cities during the 1980s. It may well reassert itself if the southern economy ever fully recovers from the recession years of the early 1990s.

## House prices

If there is one feature which typifies the overheating of the southern economy during the late 1980s, it is the massive increases in house prices and the subsequent decline and stagnation of the housing market. Information from the building societies shows the pattern quite clearly (Figure 6.6). Nationally house prices peaked in the third quarter of 1989, having more than doubled over the five-year period 1984–1989. This inflation was far more marked in the South East than elsewhere. Outside the region prices rose at a more modest rate (approximately 30 per cent) and lagged behind the pace of change in the South.

A number of authors have drawn attention to the fact that house prices and housing provision have a very profound effect on the rate and direction of labour migration (see, for example, Allen and Hamnett 1991, Champion and Fielding 1992). In a country where two-thirds of all householders now own their own homes, the cost of property has become a major consideration for those contemplating moving. House-price inflation in the South exacerbated the divisions between North and South, making it very much more difficult for those outside the region to buy in to the property market. At the same time it discouraged some from leaving the region for fear of losing a rapidly increasing capital asset (not to mention the problems of buying back in at a later stage). Admittedly there were also those who could take advantage of the differentials – selling in the South and buying very much larger and well-appointed property in the North. But, generally speaking, house-price inflation acted as a negative factor in inter-regional migration.

The situation in the 1990s is almost the reverse of the late 1980s. House prices declined first of all – and most rapidly – in the South East, whereas in the North and West the picture is more stable (Figure 6.6). The effects of declining house prices have been as limiting on population movement as rapid inflation. It has led to stagnation as people have become reluctant to sell at reduced prices. In the early 1990s the combination of relatively high mortgage rates, falling property values and recession brought the highly lucrative southern property market to a grinding halt and put a significant number of estate agents out of business. Of course, for those who had bought property many years ago and had small mortgages, this inflation was unearned wealth. But for those entering the market for the first time or making a substantial step up, the high mortgage costs proved to be crippling, leading to high levels of debt and an increase in house repossession. For many people in the region, particularly those in the lower strata of the middle-income category, house purchase has resulted in significant impoverishment – paying high mortgage costs on a depreciating asset and, in some cases, leading to the problems of negative equity where the value of property has fallen below the amount borrowed for its purchase.

It is useful to try to identify the impact that these variations in house prices have had on patterns of population movement within the South East region. The spatial variation in the value of property has created

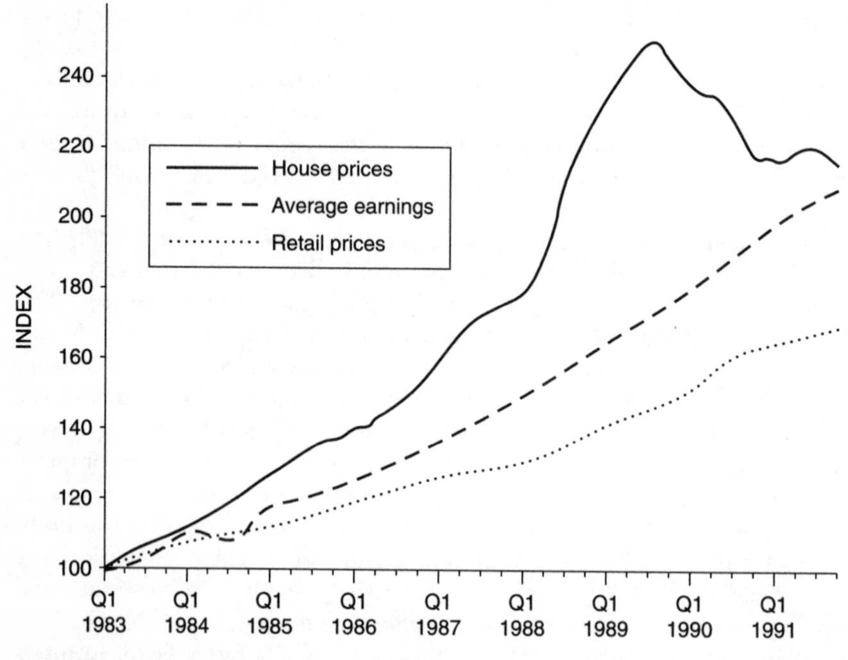


*Figure 6.6* House price trends for 1983–91 quarters and variations by region in Britain 1991

*Source:* Nationwide Building Society *House Price Index* (1992)

*Figure* 6.6 (Continued)

opportunities as well as constraints for movement at the local scale. The existence of a gradient in house prices from London to the edges of the region, for example, has provided a staircase for social mobility. By selling small but expensive houses close to London, it was possible for some people to buy larger more attractive properties in areas like Norfolk, Suffolk, Dorset or Wiltshire which were still within commuting distance of the capital. The additional costs and time of journeys to London were offset by an enhanced quality of life in pleasant rural locations. For others there was little choice. The high rates of house-price inflation were fuelled by the shortage of supply around London, and to find any affordable accommodation as a first-time buyer or in-migrant meant buying over-priced property in the capital or locating in areas that were at a distance from it. At the other end of the life-cycle, one of the factors for the outward movement on retirement has been the possibility for people to sell relatively expensive property close to their former place of work and buy cheaper property in the zones beyond the Home Counties – along the coast or further afield outside the region into the South West and Wales. The rapidly rising prices of the late 1980s encouraged a marked upturn in this longer-distance migration, with areas like central Wales and Cornwall experiencing net in-migration rates between 10 and 20 per 1,000 in 1988–9 (Stillwell *et al.* 1991).

These patterns of movement between the high-priced inner zones and lower-priced outer zones of the South are also present at a more local scale between different residential neighbourhoods. Much of the residential relocation within London has taken place over short distances, where quite sharp differences in property prices occur between adjacent areas. The variation is largely determined by the 'desirability' of areas and other locational factors. On a broader scale within the region it is possible to identify undulating house-price surfaces which reflect relatively minor changes over space but which can have an important influence on the scale and direction of more local population movements. From evidence gathered on a field study in East Kent, it is apparent that locational factors such as accessibility, geographical setting (e.g. coastal areas) and the perceived 'social tone' of places have a stronger influence on local house-prices than other factors to do with size, age or condition of property.

## Transport developments

Improvements in transport infrastructure are of key significance in the changing patterns of accessibility to London and other centres of employment in the South East. One of the more notable effects of the overcrowding in the region during the 1980s has been the inability of the existing infrastructure to cope with the higher levels of demand. The perception is that London as a city is becoming increasingly congested with traffic, that new motorway developments such as the M25 are simply not big enough to cope with levels of usage, and that the rail network is overstretched to the point where it may no longer be able to provide a reliable service to its regular users.

Nevertheless, there have been important developments in communications over recent years, and the process is ongoing, with fresh Government investment in the road programme. The M40 extension to Birmingham, the electrification of the East Coast main line, better routes to Bedford and Brighton, and eventually the high-speed link to the Channel Tunnel not only will improve connections with the rest of Europe but also with places closer to the capital that are not well served at the moment (e.g. Canterbury and Thanet). Improvements in communications can result in two principal effects. One is to extend the catchment area of London – bringing more and more places within commuting distance. But at the same time they can encourage decentralization as firms and businesses are able to locate away from the centre and still retain adequate access to the functions of the capital – as, for example, the developments along the M4 corridor.

The map of travel times by rail from London illustrates the first of these effects (Figure 6.7). It is not very precise as it only shows the high-speed Intercity routes and there are many places closer to London which are not so well served. However, it does show the spreading influence of London to the north and west along the major routes, capturing the commuter catchments of provincial centres (Bristol, Birmingham, Leicester, etc.). The map also illustrates the rather poor service for Kent and Sussex. It is perhaps noteworthy that Peterborough is only 45 minutes away from London (distance 75 miles), whereas the typical journey-time to Canterbury is 1 hour 20 minutes (50 miles from London).

Despite its very poor public image, the M25 has also had a significant impact on the patterns of accessibility around London. It has become almost a byword for congestion and from its early days has been unable to cope with the level of traffic flow on some sections. But since its opening it has led to improvements in communication between the northern and southern suburban areas, allowing for patterns of movement and commuting other than into and out of central London. It has also been of some considerable importance for the development of new peripheral industrial areas and for the location of wholesale businesses and retail parks. It has had a role to play in pushing out the zone of economic development further and further into the Home Counties and generally improving intra-regional communications (Figure 6.8).

Future patterns of population movement in the South East are likely to be affected by the continuing development of the transport infrastructure. High-speed rail links between the Channel Tunnel and the regions could influence new industrial locations and open up areas currently not that well served for commuting, for example along the North Kent coast. Similarly, improvements to the road system within the region will help to consolidate the developments around the capital and better the linkages between the increasingly dispersed elements of the region's economy. On the other hand, lack of investment might just exacerbate the current problems of congestion and act as a negative factor on further economic expansion – leaving the region at a disadvantage in relation to neighbouring regions within the European Union.

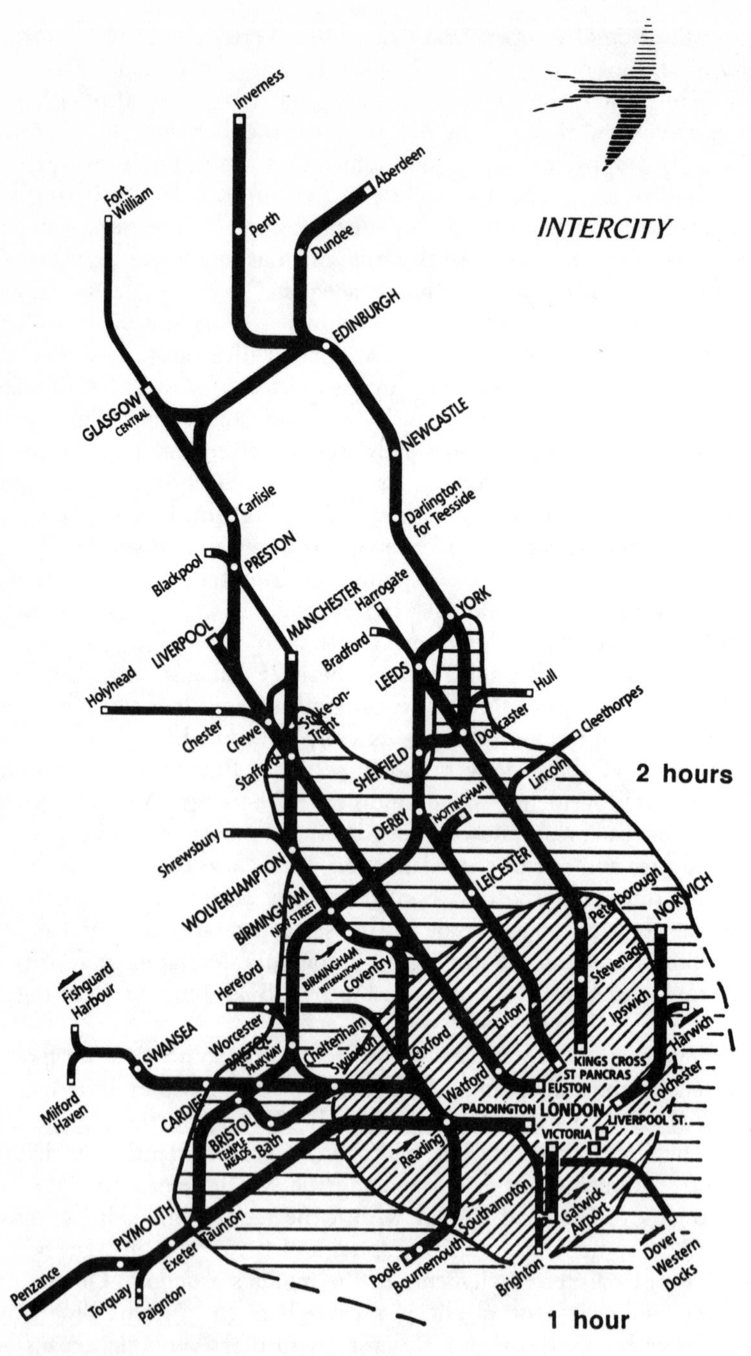


*Figure 6.7* Travel times from London 1991

*Source:* InterCity Timetable (September 1991)

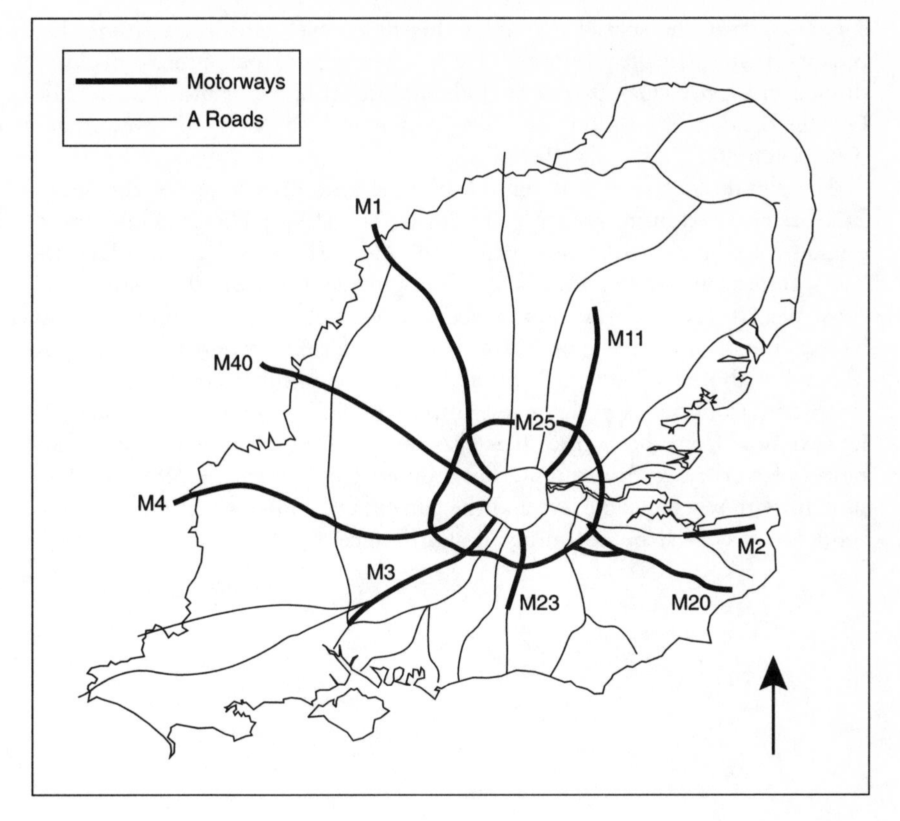


*Figure 6.8* South East communications network

*Source:* AA Road Atlas of Great Britain (1995)

## New centres of economic growth

The detailed analysis of the performance of local economies, particularly the work carried out by the Centre for Urban and Regional Development Studies at Newcastle University, has identified local labour market areas (LLMAs) which represent key growth centres in the economic restructuring of the 1980s (Champion *et al.* 1987). Not surprisingly, the ranking of these localities for the mid-1980s revealed that all the top fifteen were in the South and, with the exception of Corby, all of the bottom fifteen were in the North.

The southern localities make up an almost contiguous crescent around the north and west of London at approximately 30 miles from the centre, stretching from Bishop's Stortford to Haywards Heath. The rapid expansion of these centres has been closely associated with the sectors of the economy which did well in the 1980s – finance, business services, distribution and retailing. Subsequent work by Champion and Townsend has suggested that the spatial extent of this new development was somewhat broader than identified by the LLMAs. The prosperous subregions they discuss in *Contemporary Britain* include almost all of the South East outside Greater London, with a few notable exceptions such as the Medway towns (Champion and Townsend 1990).

The significance of this is that clearly the spatial structure of the South East has changed quite dramatically during the 1980s. The growing 'outer' economy associated with areas such as the M4 corridor and Cambridgeshire has significantly altered the focus of employment within the region and consequently lessened the dominance and centrality of London. This has had an important impact on patterns of population movement and population distribution. The tidal ebb and flow of commuters from London to the Home Counties is perhaps less important now as more employment is located away from the capital. It is also apparent that there is less of a separation of workplace from home. Distributed employment is often in close proximity to where people live and the pattern of London work and suburban residence may well be declining in significance.

# 7
# POPULATION ISSUES IN THE 1990s

In order to stress the relevance of demography as a subject for study, it is necessary to place population issues in the context of economic, social and political change in Britain in the 1990s. Population change is not an independent variable; it is influenced by underlying trends in economic development and by social and political forces. The majority of issues raised in the preceding chapters are, to an extent, socially determined. The changing patterns of marriage and family formation, health and mortality, and trends in population movement are all strongly influenced by people's conditions of life and by the opportunities available to them.

Many of these issues have also appeared on the political agenda in recent years. The increasing cost of the Welfare State has initiated a fundamental rethink of the basic principles which underlie current social provision, and questions are now being asked about whether a 'cradle to grave' system of support for all can be sustained. This may help to explain why certain groups in the population have become the focus of specific attention. For example, the experience of single parents has been highlighted as an indication of the decline in the importance of 'the family' and consequently in the stability of society. But the thinking behind this concern is as much to do with the costs to the State of supporting single-parent families as it is to do with 'moral' values.

By focusing on the life experiences of particular groups, there is an implied association between the choices made by individuals and general trends in society. A decline in moral values leads people to reject the traditions of their parents and grandparents and to adopt a less responsible approach to their role within society. However, such a view fails to appreciate the circumstances under which such decisions are made. The apparent decline in 'the family' cannot be disassociated from the changing nature of the labour market during the 1980s and 1990s and people's experience of work (or unemployment). The increasing participation of women, the growth in part-time employment and the general decline in male skilled and semi-skilled labour have all undermined the operation of the standard nuclear family and led to significant reassessments of people's social and economic roles.

In this context population can be seen as a political issue and the policies adopted by political parties will have some bearing, both directly and indirectly, on population trends in Britain. This final chapter will identify a number of these issues of concern and attempt to identify the principal directions of policy development. Hopefully this will serve to explain why a knowledge of population is essential for an understanding of society and why it is necessary to base the debate of political issues on a firm factual basis.

## Population and the world of work

There have been a number of significant changes in the size and structure of Britain's labour force over the past twenty years. The economy has undergone a transformation every bit as dramatic and undoubtedly more rapid than the transitions normally associated with the 'industrial revolution' of the early nineteenth century. The decline in traditional, labour-intensive industries in mining and manufacturing has been offset by the growth in new forms of economic activity and new patterns of work in the service industries. For many, work itself has ceased to be the central focus of life. The increase in long-term unemployment during the 1980s means that many people have in effect become excluded from the labour force and, in certain locations, generations have passed through the economically active years of the life-cycle without any realistic chance of secure employment. For many others, work has become a more irregular activity, depending on the availability of part-time employment or other casual opportunities. Even for those in more stable sectors of the economy, patterns of working have changed dramatically. Many professional people have had to face the need to develop new career directions in response to unforeseen circumstances. More people are self-employed and many have 'portfolios' of occupations, fulfilling a number of different roles.

In aggregate terms there have been a number of identifiable trends in the overall structure of the labour force. The age-profile has altered mainly as a consequence of the ups and downs in fertility in previous generations. One of the reasons why so much attention was paid to the plight of youth unemployment in the early 1980s was the sheer number of young people entering the labour market at the time. The baby boomers of the early 1960s reached the age of maturity during a period of economic recession and ran the risk of swelling the ranks of the unemployed at a time when the Government was attempting to oversee major changes in the economic structure of Britain. Similarly, ten years on, there was a dramatic fall in the number of young people as the depleted generation of the mid-1970s came of age. Between 1984 and 1994 the number of 16–19 year olds in the population fell by approximately one million. In the early 1990s this caused concern because of fears of a shortage of labour and because of longer-term concerns about the continuing trends in low fertility. As a consequence there was an attempt to attract more women into the labour force and particularly to encourage those women who had left in order to start a family to return to work as early as possible (Employment Department 1990). 'Women returners' not only helped to make up for the shortfall in young adults, but in many cases had training and experience, paid for by the State, which had only been applied for a limited period of time. For example, the National Health Service estimated that about 30,000 nurses left the service each year in the late 1980s. Several initiatives were introduced by the then Department of Employment and a number of major employers to attract women back to work, including the provision of child-care facilities. However, the concern was relatively short-lived. The recession of the early 1990s resulted in a further increase in unemployment levels and many of the worries about labour shortage evaporated. Perhaps not surprisingly, efforts to improve pre-school provision and workplace facilities for children also faded away.

There have also been significant qualitative changes in the structure of the labour market. One reason why the market has been unable to absorb the growing numbers of unemployed has been the increasing demand for employees with appropriate skills and capabilities. Gaining entry to the labour market has become more difficult as manual occupations have declined and as employers have become more selective about recruits to jobs in the service and quaternary sectors. Britain needs a better trained workforce if it is to meet the challenge from its competitors in Europe and overseas, and the current skills gap means that many find themselves excluded from the opportunities of employment. The changes that have taken place over the past twenty years were at first assumed to be the result of economic recession, with the likelihood that there would be a return towards full employment once economic conditions improved. Now it is appreciated that the assumptions that applied in the 1950s and 1960s – that the majority of the potential workforce was actually in work – no longer apply. Increasingly the labour market is becoming segmented into different groups: what Will Hutton has referred to as the 30/30/40 society, with 30 per cent of the population in permanent, well-paid employment, 30 per cent in less formal types of employment (part-time, self-employed, short-term contracts etc.) and 40 per cent in effect excluded from the labour market through lack of qualifications or appropriate skills (*Guardian* 1995a).

The third major change to the labour market over recent years has been the changing balance between the sexes. There has been a significant increase in the participation of women. Ninety per cent of the growth in employment between 1971 and 1991 was accounted for by female employment (*Employment Gazette* 1991). In 1971 women made up 37.5 per cent of the labour force. In 1991 this proportion had risen to 43.4 per cent and is expected to reach 45 per cent by the year 2001. Figure 7.1 indicates the increasing number of women in employment in the younger-adult categories and the declining number of males in the older categories – a consequence of early retirement and the displacement of unskilled and semi-skilled employees. Although far more women are now in employment, much of this work falls into the less formal sector. It represents part-time or short-term employment and 'flexible' patterns of working. Women tend to drift in and out of the labour market more than men, and employment opportunities tend to reflect conditions in

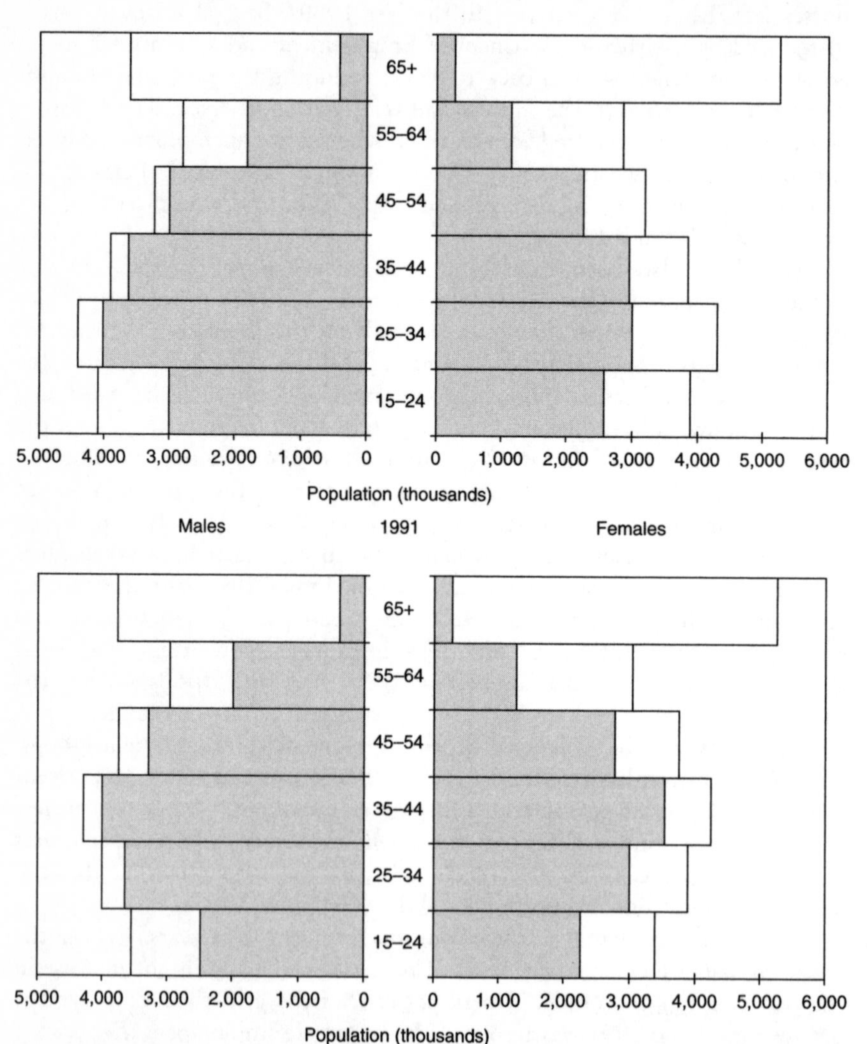


*Figure 7.1* Age/sex structure of the labour force 1991 and 2001 (shading shows numbers in employment)

*Source: Employment Gazette* (May 1991)

local economies. In her study of the changing labour market in South Wales, Doreen Massey identified the difference between a former labour market, that had developed around the primary-sector industries of mining and steel manufacture and which was heavily male-dominated with a high proportion of manual and semi-skilled labour, and a new labour market that had grown up with economic restructuring. This was typified by new jobs in the electronics industries and high levels of female employment. The former had created a patriarchal society that had remained relatively stable over many generations. The latter offered less stability and less security, and led to changes in social patterns and family organization (Massey 1984).

## The future of the family

Changing employment opportunities for women have been one of the key factors responsible for the changing patterns of family life in Britain in the 1990s. The traditional nuclear family was based on the principle of the divided roles of husband and wife, the male 'breadwinner' and the female 'carer'. Such patterns no longer apply within the more complex labour market of the 1990s and in their place a wider variety of family types has developed. Seen from a longer-term perspective, there may not be anything very surprising about this. In the past, family structure was always related to the nature of work and the structure of communities. In pre-industrial times, when family and household economies depended on contributions from all members of the family group, it was common for people to live in 'extended' families, encompassing more than one generation (children, parents, grandparents), or with lateral extensions (brothers, sisters, aunts, uncles, cousins). These groups displayed a degree of self-sufficiency and control over patterns of work. The adults and elder children worked and the very young and elderly received support from within the group. Such patterns were flexible and able to accommodate fluctuations in the amount of work available while providing a degree of stability and security.

The nuclear family (husband, wife and children) becomes of more significance during the process of industrialization. The new patterns of work determined by employment in manufacturing introduced a more regularized labour market determined by wages and defined patterns of working. Those without employment were unable to contribute to a family economy and increasingly the emphasis was placed on the wage-earners to support the whole family unit. In such circumstances 'dependents' became more of a burden to the family. The smaller nuclear family was well attuned to the needs of the industrial economy and with it developed a whole set of norms and assumptions about family life and family structure. When, in the 1990s, people reflect on what might have been lost as a result of the 'decline of the family' in recent years, the normal family that is referred to is the nuclear family and not the less formal patterns of family life associated with earlier generations.

Post-industrial, or post-modern, families of the 1990s reflect the underlying realities of the social and economic conditions of the time. The decline of marriage, the increase in divorce, the delay in family formation, the increasing number of single parents and the spread of multiple families (re-formed family groups resulting from second marriage or new associations), are all indicative of new ways of life (see Chapter 5). The concern expressed by many is that 'the breakdown of the nuclear family produces dysfunctional, under-achieving children' (*Guardian* 1995b) and that the next generation will have less of a sense of moral and social responsibility. Children who have not known a stable family upbringing are less likely themselves to form secure relationships. The consequence is the gradual undermining of social cohesion, moral degradation, lawlessness and anarchy. Lack of family support for children creates a downward spiral of poor educational performance, lack of motivation and limited opportunities for career development.

Much of the attention of these views is focused on single-parent families. Lone mothers have been transformed from an issue of social concern, a disadvantaged group who need special support, to the 'villains of the piece'. In a now famous speech by John Redwood, then Secretary of State for Wales, about single mothers on the St Mellons estate in Cardiff, he suggested that lone mothers were acting irresponsibly and that the absent fathers should return, not only to provide the financial assistance for bringing up the children but also to offer 'the normal love and support that fathers have offered down the ages' (*Guardian* 1993). The reaction by the Government in the early 1990s was to look for ways of making single parenthood less attractive. This included cutting social security benefits, removing privileges in terms of access to housing, and exploring ways of making the parents of lone mothers financially liable – all with the primary purpose of reducing the burden on public expenditure (*Guardian* 1993b). The Labour Party adopted a less punitive stance and proposed measures for encouraging lone parents to find work as an alternative to claiming benefit. These included the provision of careers advice, child-care facilities and after-school clubs. The approach is different but the primary objective is the same – to reduce welfare costs.

The financial considerations have become associated with the moral arguments; unstable or incomplete families are seen as a drain on the State and a threat to the stability of society. However, rather less is heard from single parents themselves about the reasons for choosing a different type of family unit, or about the difficulties they experience. Most single-parent families result from divorce and many lone mothers are in the older age-categories. It is not by any means exclusively a young teenage experience (see Chapter 5). What is often forgotten in the debate is that for many the experience of being brought up in a nuclear family was far from ideal. It could be restrictive and isolating – a regime of rules, regulations, controls and inhibitions. In part the move towards other types of family structure could be seen as a rejection of the experience that the current generation of young adults had as children.

It is assumed that the small independent family unit is the building block of society. But as has been shown by Young and Willmott, and others, the stability of social life is more a result of community structure than family structure. In their classic book, *Family and Kinship in East London*, they demonstrated that it was the inter-generational links, particularly between mothers and daughters, which provided the reinforcement in working-class community structure, with additional strong ties through kinship, working relations and communal life (Young and Willmott 1962). It is perhaps as much the reshaping of communities as the change in family life which has influenced contemporary perceptions of the decline of the family.

Changes in the form of family and household organization will have important consequences on people's living arrangements over the next twenty years and the demand for different types of accommodation. The total number of households in England is projected to increase by 23 per cent from 19.2 million in 1991 to 23.6 million by 2016 – an extra 4.4 million units – mainly as a result of the break-up of families, the decline in marriage and the increasing proportion of elderly within the community. The largest single sector is one-person households, with a projected growth of 3.5 million (Department of the Environment 1995). The nuclear family of previous generations was accommodated in the Victorian terraces, inter-war semi-detached properties and post-war housing-estate developments. In the future there will be a need not only for significantly more properties than at present – with consequent demands on space and conflicts over land-use – but also for different types of accommodation more in line with changing structures and life-styles. By 2016 the average household size will have fallen to only 2.17 people per household.

## Britain's ageing population

It is clear from the current trends in population change that within the next thirty years the balance between the age categories will shift significantly towards the elderly. The consequence of relatively high fertility in the early 1950s, and the baby boom of the 1960s followed by a long and sustained period of low fertility, is a bulge generation which will reach retirement age from the year 2010 onwards (Chapter 4). In addition, the improving conditions of life and health-care support for the elderly are helping to prolong the average lifespan. Life-expectancy from age 70 has increased by an average of an additional 2.1 years for men and 2.8 years for women since 1961 (*Population Trends* 87, table 12). In the next century there will not only be more elderly in total but they will also be living longer lives. The number of people over retirement age is expected to rise from 10.4 million in 1991 to approximately 13.5 million in 2030. Although there is current concern about the increasing numbers of the elderly, it is not an immediate issue. The generation that is currently aged 60 and above is the generation born during the depression years of the 1930s, a time of low fertility. The care of the elderly will be a more significant issue once the numbers increase rapidly.

An ageing population presents a number of problems that need to be addressed by government. On the one hand there is the financial problem of finding sufficient resources to secure pension payments, and on the other there is the difficulty of providing adequate support for the 'frail' elderly in the later stages of the life-cycle. The pensions issue has gathered momentum during the course of the 1990s, with the growing realization that current arrangements will not be able to cope with the increasing demands of an ageing population. As the balance between the age categories changes, there will no longer be sufficient people in work and paying tax to fund the costs of the dependent elderly. Britain's system of state pensions was introduced by the Liberal administration in the early years of the twentieth century at a time when fertility was significantly higher than today and when life-expectancy was significantly lower. It is a 'pay-as-you-go' system. The national insurance contributions of those in work pay for the pensions of the retired. This was fine during times of near full-employment, and when the retired drew pensions for a relatively short period after retiring. In 1950 there was one pensioner for every five people of working age; by 2030 it is estimated that there will be three to every five. To continue to fund pensions in the future it will mean that either the size of individual pensions will have to be reduced or those in work will need to pay higher taxes. Neither of these is particularly acceptable. Reducing the level of pensions is unfair on current generations who have paid national insurance contributions throughout their working life with the expectation of receiving the benefits of those payments on retirement. However, raising taxes is potentially very unpopular – particularly for political parties that have made a point of promising a reduction in the overall tax burden.

A third alternative is to abandon state pensions altogether and replace them with compulsory private pension funds. The scheme proposed by the Conservative Party in 1997 involves removal of entitlement to the state pension to those under the age of 25 who instead would pay into private funds through a rebate on national insurance contributions. In effect this generation would pay for pensions twice – for current pensioners through tax and national insurance, and for their own pensions through compulsory contributions to privately managed funds. It has been estimated that a national insurance rebate of £9.00 per week, paid throughout a working life, would yield an average pension of £175.00 per week at current values. The scheme proposes a state guarantee to underwrite the funds and protect against the vagaries of stock-market fluctuations.

The scheme attempts to address the issue of finding additional resources for pensions without increasing the burden on the State. However, a number of commentators have pointed out significant flaws in the proposals, including an over-optimistic assessment of the returns from private funds and the overall increase in the tax burden brought about by the removal of tax exemption on all pension contributions. The scheme also fails to accommodate the changes to the labour market identified earlier. It assumes the accumulation of a private fund during a long working life. For those with gaps in their employment record, including many women and self-employed, it will be difficult to accumulate sufficient funds to provide an adequate pension. The promises of the State to underwrite the scheme to the value of the present basic pension – uprated by price inflation – will only go some way to resolving the issue. By breaking the link with earnings under SERPS (State Earnings Related Pension Scheme) in the early 1990s, and replacing it with a link with prices, the value of the basic pension has declined. Those who have been unable to maintain a regular income during their working life will, in all probability, face continuing problems on retirement. In addition, there are doubts about the ability of the private sector to manage such a large and complex scheme. Experience in the 1980s, when people were encouraged to opt out of company schemes and buy personal pensions, resulted in over 600,000 people being worse off largely as a result of poorly regulated trading. A privately managed scheme may be more difficult to oversee and regulate than current publicly funded arrangements.

Such concerns are part of a wider debate about the ability of the State to provide universal support for all. Another solution to the problem would be to provide state pensions only for those who were in greatest need, based on a system of 'means testing'. This might be difficult to justify to those who had spent a working life paying national insurance contributions and who saw no apparent return on this investment, although most might accept that national insurance was, in reality, simply another form of taxation.

## Care of the elderly

This same dilemma applies to the State's responsibility for the care of the elderly. Support post-retirement does not simply extend to the provision of pensions. With increasing age there is an increasing demand for medical care and other support services. Forty per cent of the retired population in Britain are over the age of 75. With improvements in life-expectancy the numbers of these 'frail' elderly will continue to expand although, as noted earlier, the major increase will not occur until the second decade of the twenty-first century. Fifty per cent of all health service beds are occupied by people over the age of 75 (Thane 1989). The annual cost of long-term care for the elderly and the chronically sick is estimated at more than £6.5

billion (*Financial Times* 1992). Over recent years there has been an attempt to shift the long-stay elderly from the geriatric wards of the newly established hospital trusts into other forms of residential accommodation. The National Health Service and Care in the Community Act of 1993 gave responsibility to local authorities to assess and provide for the care needs of the elderly. As far as possible many authorities have sought to keep down costs by encouraging the elderly to stay in their own homes – supported by visiting care staff. In addition, the legislation requires the elderly to use their own resources, where available, to pay for residential care. Currently those with assets in excess of £16,000 are required to pay. Once this threshold is crossed the local authority will step in to pick up the bill, although without guaranteeing continuity of residence. Those who pay for expensive care may be transferred to less expensive establishments. In effect this has required many of the elderly to sell their homes to fund their own care – an issue which has become of some political significance as the wealth generated during a working life-time is spent on care rather than being passed on to the next generation.

One in four of the population aged over the age of 85 lives in an institution of some kind, either a nursing home with professional care or a residential home. The numbers in care are set to increase by 100,000 or more over the next ten years. At the same time the support for the elderly within families is declining. The role of daughters as informal carers is much less common today than twenty years ago. More women are in work or are raising families later in the life-cycle. In some cases it is now the 'young elderly' who accept the role of looking after the 'old elderly' – women who have retired with surviving parents (usually mothers) twenty-five to thirty years older.

Within the private sector, care for the elderly is becoming big business. The cost of residential care ranges between £1,000 and £2,000 per person each month. The market is underprovided for at present and is set to grow steadily over the foreseeable future. The public sector is currently looking for ways of reducing capacity in hospitals, and in older, expensive, local authority-run homes. In addition, there are guarantees that the State will underwrite the cost of care for those who cannot afford it. Whether these arrangements will be sufficient for the larger numbers of elderly in the population in the future is a matter of some concern. As with the dilemma over pensions, the Government is faced with a limited range of choices: it could raise taxes to increase the amount of state support for the care of the elderly; it could encourage (or require) individuals to fund their own care through private insurance schemes, or it could increase the age of retirement to keep people in the labour market for a longer period of time. This latter proposal is probably unacceptable on social and political grounds. Higher taxes will be difficult to justify, but a dependence on private insurance for both pensions and residential care is placing a heavy burden on the market to generate sufficient returns and leaves people vulnerable to the vagaries of the stock market. In addition, people will need to pay premiums over a considerable period of time to generate sufficient capital to pay for the spiralling costs of care. As always the choices in politics are not easy.

## Britain's population in the twenty-first century

The theme of this book is that population issues are important and influence many aspects of social, economic and political life in Britain. Predicting the future trends in population in the next century is a risky business. As indicated in Chapter 4, attempts to estimate trends in population in earlier decades proved to be wildly inaccurate. In the mid-1950s nobody expected the rapid upturn in fertility that occurred over the following ten years, and the consequences of that fairly sudden change in demographic behavior have left reverberations in the country's age structure which will go on influencing social issues into the next century. On the other hand, it is important to have a clear idea about future trends in order to plan long term for the allocation of state resources and cater for changing levels of demand in social provision such as health care, education and housing.

In simple terms it is possible to identify three possible scenarios: a stable population; population decline; and a return to population growth. Of these the first is the easiest to contemplate and presents least problems in terms of forward planning. The population in Britain has remained relatively stable over the past thirty years. Between 1971 and 1991 the total population increased by 1.8 million (3.3 per cent). Total live births have fallen from 869,900 (1971) to 708,900 (1996). The total number of deaths has remained fairly constant at around 600,000. Fertility levels have also remained stable with a TPFR (total period fertility rate) of between 1.66 and 1.82. A stable regime of this kind would maintain the population at its present size with relatively minor ups and downs in fertility reflecting the varying sizes of different age-groups. The current levels of fertility are below long-term replacement rates but have not yet led to a decline in numbers because of improvements in life-expectancy and the current imbalance in numbers (a relatively small elderly age-group and a relatively large middle-age category). Over time this situation will change and crude death rates will rise when the 1960s cohort reaches the final stages of the life-cycle.

This may be the time when the second scenario, population decline, becomes more significant. If fertility rates remain below replacement indefinitely (TPFR 2.1), eventually the total size of the population will decrease. The prospect of population decline raises many of the issues that were first discussed back in the 1930s (Chapter 4). It would have the positive benefit of reducing the population to the level of available resources and employment, and may add to the overall quality of life in Britain. But on the negative side it would exacerbate the difficulties of managing an ageing population. The larger elderly age-categories would not be matched by the younger groups and the difficulties of raising sufficient tax and resources indicated in the last section would become more problematic. It may also in the long run reduce the overall status of Britain – although the trends in Britain are currently in line with other European countries and it is likely that population decline would be a common feature of much of the developed world.

The scenario of population growth appears the least likely. The changes in social behaviour identified over the past twenty years, in particular the decline of marriage and the shift of family formation to later in the life-cycle, would suggest that the potential for an upturn in fertility is limited unless there is a significant reversal of current conditions. One of the major factors behind recent changes in fertility has been the participation of women in the workplace. It is hard to imagine a wholesale return to the social conditions of the 1950s and 1960s, or to former patterns of family life. Nevertheless, an increase in population cannot be ruled out. The underlying influence on the trends in fertility has been the performance of the national economy. A sustained improvement in economic conditions might contribute, once again, to younger patterns of family formation and changes in desired family size. An increasing population may in itself contribute to economic expansion and reduce the problems of dependency in the longer term. However, in the short term it would fairly rapidly lead to an increase in the number of children in the population and renew pressure on government for increasing educational provision and other services.

Clearly, trends in population change have an impact on policy formation. The question remains as to whether policy formation can influence the dynamics of population in Britain and whether government should be seeking to intervene to influence population change. There are a number of areas where policy could make an impact through the provision of services or through the manipulation of taxation. The removal of tax relief on mortgages, for instance, would increase the cost of house buying and add pressure on couples to delay starting families. The allocation of resources to nursery education would encourage more women to stay in work and raise families. Proposals from the Conservative Party to allow a partner's tax allowance to be claimed if one of the parents remains at home to bring up children are, in part, designed to encourage stable families. Legislation to tighten up on abortion would influence trends in fertility. Restrictive legislation on smoking – or increasingly punitive taxation – would in time lead to a reduction in the number of smoking-related deaths. In these ways, although government may not espouse a 'population policy', it is able to influence the dynamics of population change. It is necessary to have a clear understanding of population issues in order to formulate policy which will improve the quality of life in Britain.

# APPENDIX: STATISTICAL TABLES

*Table A* Residents by age 1991

| *Area* | *Total persons* | *Percentage aged:* | | | | | | | | | | | *Percentage of all persons with limiting long-term illness* |
|---|---|---|---|---|---|---|---|---|---|---|---|---|---|
| | | *0–4* | *5–15* | *16–17* | *18–29* | *30–44* | *45 to pen.*[a] | *pen.*[a] *to 74* | *75–84* | *85+* | *75+ Males* | *75+ Females* | |
| Great Britain | 54,888,844 | 6.6 | 13.5 | 2.5 | 18.2 | 21.2 | 19.3 | 11.7 | 5.5 | 1.5 | 2.4 | 4.6 | 13.1 |
| England and Wales | 49,890,277 | 6.6 | 13.4 | 2.5 | 18.2 | 21.2 | 19.2 | 11.7 | 5.6 | 1.5 | 2.4 | 4.7 | 13.1 |
| England | 47,055,204 | 6.7 | 13.4 | 2.5 | 18.3 | 21.3 | 19.2 | 11.6 | 5.6 | 1.5 | 2.4 | 4.6 | 12.8 |
| Regions of England | | | | | | | | | | | | | |
| North | 3,026,732 | 6.5 | 13.8 | 2.5 | 17.4 | 21.2 | 19.7 | 12.4 | 5.2 | 1.4 | 2.2 | 4.4 | 15.8 |
| Tyne and Wear | 1,095,152 | 6.5 | 13.5 | 2.4 | 17.9 | 21.0 | 19.2 | 12.7 | 5.3 | 1.4 | 2.2 | 4.5 | 16.7 |
| Remainder | 1,931,580 | 6.4 | 13.9 | 2.6 | 17.1 | 21.3 | 19.9 | 12.2 | 5.2 | 1.3 | 2.2 | 4.3 | 15.3 |
| Yorks and Humberside | 4,836,524 | 6.7 | 13.6 | 2.6 | 18.1 | 20.9 | 19.2 | 11.9 | 5.5 | 1.5 | 2.4 | 4.6 | 14.2 |
| South Yorks | 1,262,630 | 6.6 | 13.1 | 2.5 | 18.6 | 20.7 | 19.4 | 12.2 | 5.5 | 1.4 | 2.4 | 4.5 | 16.6 |
| West Yorks | 2,013,693 | 7.1 | 14.2 | 2.6 | 18.5 | 21.1 | 18.6 | 11.3 | 5.3 | 1.4 | 2.2 | 4.5 | 13.7 |
| Remainder | 1,560,201 | 6.4 | 13.3 | 2.6 | 17.2 | 21.0 | 19.8 | 12.3 | 5.8 | 1.6 | 2.6 | 4.9 | 13.0 |
| East Midlands | 3,953,372 | 6.6 | 13.6 | 2.6 | 17.9 | 21.4 | 19.5 | 11.7 | 5.3 | 1.4 | 2.4 | 4.3 | 12.6 |
| East Anglia | 2,027,004 | 6.3 | 13.3 | 2.6 | 17.4 | 21.2 | 19.3 | 12.2 | 6.0 | 1.7 | 2.8 | 4.8 | 11.7 |
| South East | 17,208,264 | 6.7 | 13.0 | 2.4 | 19.2 | 22.0 | 18.8 | 10.9 | 5.5 | 1.6 | 2.4 | 4.6 | 11.3 |
| Greater London | 6,679,699 | 6.9 | 12.6 | 2.2 | 21.6 | 22.2 | 17.7 | 10.2 | 5.2 | 1.4 | 2.2 | 4.4 | 11.8 |
| Inner London | 2,504,451 | 7.1 | 12.1 | 2.0 | 24.2 | 22.4 | 16.6 | 9.6 | 4.8 | 1.3 | 2.0 | 4.0 | 12.7 |
| Outer London | 4,175,248 | 6.8 | 12.9 | 2.3 | 19.9 | 22.1 | 18.4 | 10.5 | 5.4 | 1.5 | 2.4 | 4.6 | 11.2 |
| Outer Metropolitan | 5,544,607 | 6.6 | 13.5 | 2.6 | 18.2 | 22.3 | 19.8 | 10.5 | 5.0 | 1.4 | 2.2 | 4.2 | 10.0 |
| Outer South East | 4,983,958 | 6.4 | 13.1 | 2.5 | 17.3 | 21.2 | 19.0 | 12.2 | 6.4 | 1.9 | 2.9 | 5.5 | 12.0 |

*Table A* (Continued)

| *Area* | *Total persons* | *Percentage aged:* | | | | | | | | | | | *Percentage of all persons with limiting long-term illness* |
|---|---|---|---|---|---|---|---|---|---|---|---|---|---|
| | | *0–4* | *5–15* | *16–17* | *18–29* | *30–44* | *45 to pen.*[a] | *pen.*[a] *to 74* | *75–84* | *85+* | *75+ Males* | *75+ Females* | |
| South West | 4,609,424 | 6.2 | 12.7 | 2.5 | 16.8 | 20.6 | 19.6 | 13.1 | 6.7 | 1.9 | 3.0 | 5.6 | 12.7 |
| West Midlands | 5,150,187 | 6.8 | 13.9 | 2.6 | 18.1 | 20.7 | 19.8 | 11.7 | 5.1 | 1.3 | 2.2 | 4.2 | 13.0 |
| Metropolitan | 2,551,671 | 7.2 | 14.2 | 2.6 | 18.9 | 19.8 | 19.2 | 11.8 | 5.2 | 1.3 | 2.2 | 4.3 | 13.7 |
| Remainder | 2,598,516 | 6.5 | 13.6 | 2.6 | 17.3 | 21.6 | 20.4 | 11.5 | 5.1 | 1.3 | 2.3 | 4.2 | 12.4 |
| North West | 6,243,697 | 6.9 | 14.1 | 2.6 | 17.9 | 20.7 | 19.2 | 11.7 | 5.5 | 1.4 | 2.2 | 4.7 | 15.0 |
| Greater Manchester | 2,499,441 | 7.2 | 14.2 | 2.6 | 18.6 | 20.7 | 18.7 | 11.3 | 5.3 | 1.4 | 2.2 | 4.5 | 15.0 |
| Merseyside | 1,403,642 | 6.9 | 14.2 | 2.6 | 17.7 | 20.2 | 19.2 | 12.1 | 5.5 | 1.5 | 2.2 | 4.8 | 16.1 |
| Remainder | 2,340,614 | 6.7 | 13.8 | 2.6 | 17.2 | 21.0 | 19.8 | 11.8 | 5.6 | 1.5 | 2.4 | 4.7 | 14.2 |
| Wales | 2,835,073 | 6.6 | 13.8 | 2.6 | 16.9 | 20.3 | 19.7 | 12.8 | 5.8 | 1.5 | 2.5 | 4.8 | 17.1 |
| Scotland | 4,998,567 | 6.3 | 13.9 | 2.6 | 18.1 | 21.4 | 19.5 | 11.6 | 5.2 | 1.4 | 2.1 | 4.4 | 13.7 |

*Source:* OPCS (1992), table E



*Note: a* = pensioner age

*Table B* Ethnic group of residents 1991

| *Area* | *Total persons* | *Ethnic group – percentage* | | | | | | | | | |
|---|---|---|---|---|---|---|---|---|---|---|---|
| | | *White* | *Black Carib.* | *Black African* | *Black other* | *Indian* | *Pakistani* | *Bangla-deshi* | *Chinese* | *Other groups* | |
| | | | | | | | | | | *Asian* | *Other* |
| Great Britain | 54,888,844 | 94.5 | 0.9 | 0.4 | 0.3 | 1.5 | 0.9 | 0.3 | 0.3 | 0.4 | 0.5 |
| England and Wales | 49,890,277 | 94.1 | 1.0 | 0.4 | 0.4 | 1.7 | 0.9 | 0.3 | 0.3 | 0.4 | 0.6 |
| England | 47,055,204 | 93.8 | 1.1 | 0.4 | 0.4 | 1.8 | 1.0 | 0.3 | 0.3 | 0.4 | 0.6 |
| Regions of England | | | | | | | | | | | |
| North | 3,026,732 | 98.7 | 0.0 | 0.0 | 0.1 | 0.3 | 0.3 | 0.1 | 0.2 | 0.1 | 0.2 |
| Tyne and Wear | 1,095,152 | 98.2 | 0.0 | 0.1 | 0.1 | 0.4 | 0.3 | 0.3 | 0.3 | 0.1 | 0.2 |
| Remainder | 1,931,580 | 99.0 | 0.0 | 0.0 | 0.1 | 0.2 | 0.3 | 0.0 | 0.1 | 0.1 | 0.1 |
| Yorks and Humberside | 4,836,524 | 95.6 | 0.4 | 0.1 | 0.2 | 0.8 | 2.0 | 0.2 | 0.2 | 0.2 | 0.4 |
| South Yorks | 1,262,630 | 97.1 | 0.5 | 0.1 | 0.2 | 0.3 | 1.0 | 0.1 | 0.2 | 0.1 | 0.4 |
| West Yorks | 2,013,693 | 91.8 | 0.7 | 0.1 | 0.3 | 1.7 | 4.0 | 0.3 | 0.2 | 0.2 | 0.5 |
| Remainder | 1,560,201 | 99.1 | 0.0 | 0.1 | 0.1 | 0.2 | 0.1 | 0.1 | 0.1 | 0.1 | 0.2 |
| East Midlands | 3,953,372 | 95.2 | 0.6 | 0.1 | 0.3 | 2.5 | 0.4 | 0.1 | 0.2 | 0.2 | 0.4 |
| East Anglia | 2,027,004 | 97.9 | 0.2 | 0.1 | 0.4 | 0.3 | 0.3 | 0.1 | 0.2 | 0.2 | 0.4 |
| South East | 17,208,264 | 90.1 | 1.9 | 1.0 | 0.6 | 2.6 | 0.8 | 0.6 | 0.5 | 0.8 | 1.0 |
| Greater London | 6,679,699 | 79.8 | 4.4 | 2.4 | 1.2 | 5.2 | 1.3 | 1.3 | 0.8 | 1.7 | 1.8 |
| Inner London | 2,504,451 | 74.4 | 7.1 | 4.4 | 2.0 | 3.0 | 1.2 | 2.8 | 1.1 | 1.8 | 2.3 |
| Outer London | 4,175,248 | 83.1 | 2.7 | 1.3 | 0.7 | 6.5 | 1.4 | 0.4 | 0.7 | 1.6 | 1.5 |
| Outer Metropolitan | 5,544,607 | 95.7 | 0.5 | 0.1 | 0.2 | 1.3 | 0.8 | 0.2 | 0.3 | 0.4 | 0.5 |
| Outer South East | 4,983,958 | 97.8 | 0.3 | 0.1 | 0.2 | 0.5 | 0.2 | 0.1 | 0.2 | 0.2 | 0.4 |
| South West | 4,609,424 | 98.6 | 0.3 | 0.1 | 0.1 | 0.2 | 0.1 | 0.1 | 0.1 | 0.1 | 0.3 |

*Table B* (Continued)

| *Area* | *Total persons* | *Ethnic group – percentage* | | | | | | | | | |
|---|---|---|---|---|---|---|---|---|---|---|---|
| | | *White* | *Black Carib.* | *Black African* | *Black other* | *Indian* | *Pakistani* | *Bangla-deshi* | *Chinese* | *Other groups* | |
| | | | | | | | | | | *Asian* | *Other* |
| West Midlands | 5,150,187 | 98.6 | 0.3 | 0.1 | 0.1 | 0.2 | 0.1 | 0.1 | 0.1 | 0.1 | 0.3 |
| Metropolitan | 2,551,671 | 85.4 | 2.8 | 0.2 | 0.6 | 5.5 | 3.5 | 0.7 | 0.2 | 0.3 | 0.7 |
| Remainder | 2,598,516 | 98.0 | 0.2 | 0.0 | 0.1 | 0.7 | 0.4 | 0.1 | 0.1 | 0.1 | 0.2 |
| North West | 6,243,697 | 96.1 | 0.3 | 0.1 | 0.3 | 0.9 | 1.2 | 0.2 | 0.3 | 0.1 | 0.4 |
| Greater Manchester | 2,499,441 | 94.1 | 0.7 | 0.2 | 0.4 | 1.2 | 2.0 | 0.5 | 0.3 | 0.2 | 0.5 |
| Merseyside | 1,403,642 | 98.2 | 0.2 | 0.2 | 0.3 | 0.2 | 0.1 | 0.1 | 0.4 | 0.1 | 0.4 |
| Remainder | 2,340,614 | 97.0 | 0.1 | 0.0 | 0.1 | 1.0 | 1.1 | 0.1 | 0.1 | 0.1 | 0.2 |
| Wales | 2,835,073 | 98.5 | 0.1 | 0.1 | 0.1 | 0.2 | 0.2 | 0.1 | 0.2 | 0.1 | 0.3 |
| Scotland | 4,998,567 | 98.7 | 0.0 | 0.1 | 0.1 | 0.2 | 0.4 | 0.0 | 0.2 | 0.1 | 0.2 |

*Source:* OPCS (1992), table J

*Table C* Vital statistics summary (Britain)

| | *All live births* | | *Live births outside marriage* | | *Marriages* | | *Divorces* | | *Deaths* | | *Infant mortality* | | *Neonatal mortality* | | *Perinatal mortality* | |
|---|---|---|---|---|---|---|---|---|---|---|---|---|---|---|---|---|
| | *No.*[a] | *Rate*[b] | *No.* | *Rate*[c] | *No.* | *Rate*[d] | *No.* | *Rate*[e] | *No.* | *Rate*[b] | *No.* | *Rate*[c] | *No.* | *Rate*[c] | *No.* | *Rate*[f] |
| *1966* | 946.4 | 17.8 | 73.2 | 77 | 426.3 | — | 42.6 | — | 627.3 | 11.8 | 18.4 | 19.4 | 12.4 | 13.1 | 25.6 | 26.6 |
| *1971* | 869.9 | 16.0 | 72.7 | 84 | 447.2 | 68.5 | 79.2 | 5.8 | 628.9 | 11.6 | 15.4 | 17.8 | 10.3 | 11.8 | 19.8 | 22.5 |
| *1976* | 649.2 | 11.9 | 59.8 | 92 | 396.1 | 57.3 | 134.8 | 9.8 | 663.8 | 12.1 | 9.30 | 14.3 | 6.33 | 9.7 | 11.7 | 17.8 |
| *1981* | 703.5 | 12.8 | 89.4 | 127 | 388.2 | 49.4 | 155.6 | 11.5 | 641.7 | 11.7 | 7.80 | 11.1 | 4.70 | 6.7 | 8.37 | 11.8 |
| *1986* | 726.8 | 13.2 | 154.9 | 213 | 383.7 | 43.4 | 166.7 | 12.7 | 644.7 | 11.7 | 6.89 | 9.5 | 3.83 | 5.3 | 7.04 | 9.6 |
| *1991* | 766.2 | 13.6 | 230.8 | 301 | 340.5 | 36.1 | 171.1 | 13.2 | 631.1 | 11.3 | 5.63 | 7.3 | 3.34 | 4.4 | 6.23 | 8.1 |
| 1993 | 736.8 | 13.0 | 236.4 | 321 | 332.6 | 35.0 | 177.8 | 13.7 | 642.8 | 11.4 | 4.65 | 6.3 | 3.05 | 4.2 | 6.64[g] | 9.0 |
| *1994* | 726.4 | 12.8 | 234.8 | 323 | 322.5 | 34.0 | 171.3 | 13.2 | 612.5 | 10.8 | 4.48 | 6.2 | 2.98 | 4.1 | 6.50[g] | 8.9 |
| *1995* | 708.2 | 12.4 | 240.2 | 339 | 313.6 | — | 167.7 | 12.9 | 626.4 | 11.0 | 4.35 | 6.1 | 2.94 | 4.1 | 6.29[g] | 8.8 |
| *1996* | 708.7 | 12.4 | 254.0 | 358 | — | — | 165.4 | — | 623.7 | 11.0 | 4.36 | 6.1 | 2.91 | 4.1 | 6.18[g] | 8.7 |

*Source: Population Trends* 89, table 8
Office for National Statistics, Crown Copyright

*Notes:*
*a* Numbers in thousands
*b* Per 1,000 population all ages
*c* Per 1,000 live births
*d* Persons marrying per 1,000 unmarried population 16 and over
*e* Per 1,000 married population
*f* Per 1,000 live and still-births
*g* Figures include still-births of 24–7 weeks gestation

*Table D* Live births: age of mother (England and Wales)

| *Year* | *Age of mother at birth* | | | | | | | *Age of mother at birth* | | | | | | | *Mean age at birth* | *TPFR*[b] |
|---|---|---|---|---|---|---|---|---|---|---|---|---|---|---|---|---|
| | *All ages* | *Under 20* | *20–24* | *25–29* | *30–34* | *35–39* | *40 and over* | *All ages* | *Under 20* | *20–24* | *25–29* | *30–34* | *35–39* | *40 and over* | | |
| | *Total live births (thousands)* | | | | | | | *Age-specific fertility rates*[a] | | | | | | | | |
| *1961* | 811.3 | 59.8 | 249.8 | 248.5 | 152.3 | 77.5 | 23.3 | 89.2 | 37.3 | 172.6 | 176.9 | 103.1 | 48.1 | 15.0 | 27.6 | 2.77 |
| *1964* (max) | 876.0 | 76.7 | 276.1 | 270.7 | 153.5 | 75.4 | 23.6 | 92.9 | 42.5 | 181.6 | 187.3 | 107.7 | 49.8 | 13.7 | 27.2 | 2.93 |
| *1966* | 849.8 | 86.7 | 285.8 | 253.7 | 136.4 | 67.0 | 20.1 | 90.5 | 47.7 | 176.0 | 174.0 | 97.3 | 45.3 | 12.5 | 26.8 | 2.75 |
| *1971* | 783.2 | 82.6 | 285.7 | 247.2 | 109.6 | 45.2 | 12.7 | 83.5 | 50.6 | 152.9 | 153.2 | 77.1 | 32.8 | 8.7 | 26.2 | 2.37 |
| *1976* | 584.3 | 57.9 | 182.2 | 220.7 | 90.8 | 26.1 | 6.5 | 60.4 | 32.2 | 109.3 | 118.7 | 57.2 | 18.6 | 4.8 | 26.4 | 1.71 |
| *1977* (min) | 569.3 | 54.5 | 174.5 | 207.9 | 100.8 | 25.5 | 6.0 | 58.1 | 29.4 | 103.7 | 117.5 | 58.6 | 18.2 | 4.4 | 26.5 | 1.66 |
| *1981* | 634.5 | 56.6 | 194.5 | 215.8 | 126.6 | 34.2 | 6.9 | 61.3 | 28.1 | 105.3 | 129.1 | 68.6 | 21.7 | 4.9 | 26.8 | 1.80 |
| *1991* | 699.2 | 52.4 | 173.4 | 248.7 | 161.3 | 53.6 | 9.8 | 63.6 | 33.0 | 89.3 | 119.4 | 86.7 | 32.1 | 5.3 | 27.7 | 1.82 |
| *1993* | 673.5 | 45.1 | 152.0 | 236.0 | 171.1 | 58.8 | 10.5 | 62.6 | 31.0 | 82.7 | 114.1 | 87.0 | 34.1 | 6.2 | 28.1 | 1.76 |
| *1994* | 664.7 | 42.0 | 140.2 | 229.1 | 179.6 | 63.1 | 10.7 | 61.9 | 29.0 | 79.4 | 112.1 | 88.7 | 35.8 | 6.4 | 28.4 | 1.75 |
| *1995* | 648.1 | 41.9 | 130.7 | 217.4 | 181.2 | 65.5 | 11.3 | 60.4 | 28.5 | 76.8 | 108.6 | 87.3 | 36.2 | 6.8 | 28.5 | 1.72 |
| *1996* | 649.5 | 44.7 | 125.7 | 211.1 | 186.4 | 69.5 | 12.1 | 60.2 | 29.6 | 77.2 | 106.8 | 88.4 | 37.0 | 7.2 | 28.6 | 1.74 |

*Source: Population Trends* 89, table 9
Office for National Statistics, Crown Copyright

*Notes:*
*a* Births per 1,000 women in the age group
*b* TPFR (total period fertility rate) is the average number of children that would be born if women experienced the age-specific fertility rates of the period in question throughout their child-bearing life-span
During the post-Second World War period the TPFR reached a maximum in 1964 and a minimum in 1977

*Table E* Live births outside marriage: age of mother and type of registration (England and Wales)

| *Year* | *Age of mother at birth* | | | | | | | *Age of mother at birth* | | | | | | *Registration*[a] | | |
|---|---|---|---|---|---|---|---|---|---|---|---|---|---|---|---|---|
| | *All ages* | *Under 20* | *0–24* | *25–29* | *30–34* | *35 and over* | *Mean age (years)* | *All ages* | *Under 20* | *20–24* | *25–29* | *30–34* | *35 and over* | *Joint* Same address[b] | Different address[b] | *Sole* |
| | *Thousands* | | | | | | | *Percentage of total births* | | | | | | *As a percentage of all births outside marriage* | | |
| *1961* | 48.5 | 11.9 | 15.5 | 9.3 | 6.2 | 5.6 | 25.40 | 6.0 | 19.9 | 6.2 | 3.7 | 4.1 | 5.5 | – | – | – |
| *1966* | 67.1 | 20.6 | 22.0 | 11.9 | 6.9 | 5.8 | 24.33 | 7.9 | 23.7 | 7.7 | 4.7 | 5.0 | 6.6 | 38.3 | | 61.7 |
| *1971* | 65.7 | 21.6 | 22.0 | 11.5 | 6.2 | 4.3 | 23.78 | 8.4 | 26.1 | 7.7 | 4.7 | 5.7 | 7.4 | 45.5 | | 54.5 |
| *1976* | 53.8 | 19.8 | 16.6 | 9.7 | 4.7 | 2.9 | 23.34 | 9.2 | 34.2 | 9.1 | 4.4 | 5.2 | 8.9 | 51.0 | | 49.0 |
| *1981* | 81.0 | 26.4 | 28.8 | 14.3 | 7.9 | 3.6 | 23.47 | 12.8 | 46.7 | 14.8 | 6.6 | 6.2 | 8.7 | 58.2 | | 41.8 |
| *1991* | 211.3 | 43.4 | 77.8 | 52.4 | 25.7 | 11.9 | 24.84 | 30.2 | 82.9 | 44.9 | 21.1 | 16.0 | 18.8 | 54.6 | 19.8 | 25.6 |
| *1992* | 215.2 | 40.1 | 77.1 | 55.9 | 28.9 | 13.3 | 25.21 | 31.2 | 83.7 | 47.2 | 22.8 | 17.3 | 19.8 | 55.4 | 20.7 | 23.9 |
| *1993* | 216.5 | 38.2 | 75.0 | 57.5 | 31.4 | 14.4 | 25.46 | 32.2 | 84.8 | 49.4 | 24.4 | 18.4 | 20.7 | 54.8 | 22.0 | 23.2 |
| *1994* | 215.5 | 35.9 | 71.0 | 58.5 | 34.0 | 16.1 | 25.80 | 32.4 | 85.5 | 50.6 | 25.5 | 18.9 | 21.8 | 57.5 | 19.8 | 22.7 |
| *1995* | 219.9 | 36.3 | 69.7 | 59.6 | 37.0 | 17.4 | 25.98 | 33.9 | 86.6 | 53.3 | 27.4 | 20.4 | 22.6 | 58.1 | 20.1 | 21.8 |
| *1996* | 232.7 | 39.3 | 71.1 | 62.3 | 40.5 | 19.4 | 26.08 | 35.8 | 88.0 | 56.5 | 29.5 | 21.7 | 23.9 | 58.1 | 19.9 | 21.9 |

*Source: Population Trends* 89, table 10
Office for National Statistics, Crown Copyright
*Notes:*
*a* Births outside marriage can be registered by both the mother and father (joint) or by the mother alone (sole)
*b* Usual address of parents

*Table F* Expectation of life at birth and selected ages (UK)

| *Year* | *Males* | | | | | | | | *Females* | | | | | | | |
|---|---|---|---|---|---|---|---|---|---|---|---|---|---|---|---|---|
| | *At birth* | *At age* 5 | *20* | *30* | *50* | *60* | *70* | *80* | *At birth* | *At age* 5 | *20* | *30* | *50* | *60* | *70* | *80* |
| *1961* | 67.9 | 64.9 | 50.4 | 40.9 | 22.6 | 15.0 | 9.3 | 5.2 | 73.8 | 70.4 | 55.7 | 46.0 | 27.4 | 19.0 | 11.7 | 6.3 |
| *1971* | 68.8 | 65.3 | 50.9 | 41.3 | 23.0 | 15.3 | 9.5 | 5.5 | 75.0 | 71.4 | 56.7 | 47.0 | 28.3 | 19.8 | 12.5 | 6.9 |
| *1981* | 70.8 | 66.9 | 52.3 | 42.7 | 24.1 | 16.3 | 10.1 | 5.7 | 76.8 | 72.7 | 57.9 | 48.1 | 29.2 | 20.8 | 13.3 | 7.5 |
| *1986* | 71.9 | 67.8 | 53.2 | 43.6 | 24.9 | 16.8 | 10.5 | 6.0 | 77.7 | 73.5 | 58.7 | 48.9 | 29.8 | 21.2 | 13.8 | 7.9 |
| *1991* | 73.2 | 68.9 | 54.2 | 44.7 | 26.0 | 17.7 | 11.1 | 6.4 | 78.8 | 74.4 | 59.5 | 49.7 | 30.7 | 21.9 | 14.4 | 8.3 |
| *1992* | 73.4 | 69.1 | 54.4 | 44.8 | 26.1 | 17.8 | 11.1 | 6.4 | 78.9 | 74.4 | 59.6 | 49.8 | 30.7 | 22.0 | 14.4 | 8.3 |
| *1993* | 73.7 | 69.3 | 54.6 | 45.1 | 26.4 | 18.0 | 11.3 | 6.5 | 79.1 | 74.6 | 59.8 | 50.0 | 30.9 | 22.1 | 14.5 | 8.4 |
| *1994* | 73.9 | 69.5 | 54.8 | 45.2 | 26.5 | 18.1 | 11.3 | 6.5 | 79.2 | 74.7 | 59.9 | 50.1 | 31.0 | 22.2 | 14.5 | 8.4 |

*Source: Population Trends* 89, table 13

*Table G* Mortality rates[a] by cause[b], health regions 1994

| *Rates per 100,000 population* | *All circulatory diseases* | | | *All respiratory diseases* | | *Cancers*[c] | *All injuries and poisonings* | | | *Other causes* | *All causes* |
|---|---|---|---|---|---|---|---|---|---|---|---|
| | *Total* | *Ischaemic heart disease* | *Cerebro-vascular disease* | *Total* | *Bronchitis and allied conditions* | | *Total* | *Road traffic accidents* | *Suicides and open verdicts* | | |
| United Kingdom | 472 | 264 | 117 | 155 | 48 | 270 | 33 | 6 | 11 | 137 | 1,067 |
| Northern and Yorkshire | 502 | 293 | 123 | 167 | 56 | 284 | 32 | 6 | 11 | 140 | 1,125 |
| Trent | 475 | 268 | 117 | 158 | 50 | 271 | 32 | 8 | 10 | 136 | 1,072 |
| Anglia and Oxford | 421 | 230 | 105 | 144 | 41 | 256 | 32 | 8 | 10 | 136 | 989 |
| North Thames | 424 | 237 | 97 | 165 | 48 | 262 | 29 | 5 | 10 | 139 | 1,019 |
| South Thames | 430 | 227 | 102 | 155 | 46 | 259 | 26 | 5 | 11 | 128 | 998 |
| South and West | 429 | 239 | 106 | 140 | 40 | 254 | 28 | 6 | 10 | 111 | 962 |
| West Midlands | 485 | 265 | 124 | 151 | 48 | 272 | 33 | 7 | 10 | 139 | 1,080 |
| North West | 520 | 301 | 116 | 178 | 62 | 286 | 33 | 6 | 11 | 136 | 1,153 |
| England | 460 | 257 | 112 | 155 | 48 | 267 | 30 | 6 | 10 | 135 | 1,048 |
| Wales | 488 | 274 | 117 | 154 | 48 | 277 | 36 | 7 | 12 | 127 | 1,082 |
| Scotland | 567 | 315 | 163 | 149 | 50 | 307 | 47 | 7 | 16 | 161 | 1,232 |
| Northern Ireland | 549 | 321 | 140 | 194 | 47 | 267 | 46 | 10 | 10 | 110 | 1,165 |

*Source: Regional Trends* 31, table 7.11



*Notes:*

*a* Adjusted for the age structure of the population

*b* Deaths at ages under 28 days, occurring in England and Wales, are not assigned to a cause of death

*c* Malignant neoplasms only

*Table H* Social class[a] of economically active population, standard regions 1995

| | *Social Class (per cent)* | | | | | | | |
|---|---|---|---|---|---|---|---|---|
| | *Professional (I)* | *Managerial and technical (II)* | *Skilled non-manual (IIIN)* | *Skilled manual (IIIM)* | *Partly skilled (IV)* | *Unskilled (V)* | *Other*[b] | *Total econ. active (thousands)* |
| United Kingdom | 5.5 | 28.6 | 21.9 | 20.6 | 14.7 | 5.5 | 3.2 | 28,426 |
| North | 4.3 | 24.9 | 20.9 | 22.6 | 16.5 | 6.9 | 3.9 | 1,416 |
| Yorkshire and Humberside | 4.5 | 25.2 | 21.3 | 23.4 | 16.2 | 6.1 | 3.3 | 2,434 |
| East Midlands | 4.4 | 26.8 | 20.4 | 22.9 | 17.3 | 5.6 | 2.5 | 2,048 |
| East Anglia | 5.0 | 27.1 | 20.2 | 20.5 | 17.8 | 5.5 | 3.9 | 1,081 |
| South East | 6.9 | 32.7 | 23.3 | 17.5 | 12.1 | 4.5 | 3.0 | 9,059 |
| Greater London | 7.2 | 35.0 | 23.0 | 16.0 | 11.0 | 4.1 | 3.7 | 3,476 |
| Rest of South East | 6.7 | 31.3 | 23.5 | 18.5 | 12.8 | 4.7 | 2.6 | 5,583 |
| South West | 5.3 | 28.2 | 21.3 | 20.4 | 15.7 | 5.5 | 3.5 | 2,374 |
| West Midlands | 4.4 | 26.8 | 20.1 | 23.5 | 16.6 | 5.6 | 3.0 | 2,579 |
| North West | 5.3 | 27.0 | 22.9 | 21.0 | 15.5 | 5.4 | 2.9 | 2,938 |
| England | 5.6 | 28.9 | 22.0 | 20.4 | 14.7 | 5.3 | 3.1 | 23,930 |
| Wales | 5.1 | 27.0 | 20.3 | 22.4 | 15.5 | 6.5 | 3.3 | 1,304 |
| Scotland | 5.4 | 26.4 | 22.7 | 20.9 | 14.5 | 6.8 | 3.3 | 2,492 |
| Northern Ireland | 3.3 | 26.5 | 21.3 | 23.8 | 13.2 | 6.6 | 5.2 | 700 |

*Source: Regional Trends* 31, table 3.15
Office for National Statistics, Crown Copyright

*Note:*
*a* Based on occupation
*b* Includes members of the armed forces, those who did not state their occupation, and, for the unemployed, those whose previous occupation was more than eight years ago, or those who never had a job

*Table 1* International migration: age and sex (UK)[a]

| *(thousands)* | *All ages* | | | *Aged 0–14* | | | *Aged 15–24* | | | *Aged 25–44* | | | *Aged 45 and over* | | |
|---|---|---|---|---|---|---|---|---|---|---|---|---|---|---|---|
| | *Persons* | *Males* | *Females* | *Persons* | *Males* | *Females* | *Persons* | *Males* | *Females* | *Persons* | *Males* | *Females* | *Persons* | *Males* | *Females* |
| *Inflow* | | | | | | | | | | | | | | | |
| *1971* | 200 | 103 | 97 | 33 | 17 | 17 | 65 | 28 | 37 | 81 | 48 | 33 | 21 | 10 | 11 |
| *1976* | 191 | 100 | 91 | 32 | 16 | 17 | 64 | 32 | 32 | 77 | 43 | 34 | 18 | 9 | 9 |
| *1981* | 153 | 83 | 71 | 30 | 16 | 14 | 48 | 24 | 24 | 60 | 34 | 26 | 15 | 9 | 7 |
| *1986* | 250 | 120 | 130 | 45 | 22 | 23 | 79 | 34 | 45 | 101 | 49 | 51 | 25 | 16 | 10 |
| *1991* | 267 | 122 | 144 | 48 | 20 | 28 | 83 | 36 | 47 | 109 | 54 | 55 | 27 | 12 | 15 |
| *1992* | 216 | 99 | 117 | 33 | 17 | 16 | 66 | 25 | 41 | 91 | 44 | 48 | 26 | 14 | 12 |
| *1993* | 213 | 101 | 112 | 34 | 17 | 17 | 73 | 28 | 44 | 87 | 44 | 43 | 20 | 12 | 8 |
| *1994* | 253 | 126 | 127 | 36 | 22 | 14 | 76 | 30 | 47 | 117 | 60 | 57 | 24 | 15 | 9 |
| *1995* | 245 | 130 | 115 | 28 | 20 | 9 | 88 | 40 | 48 | 107 | 57 | 50 | 22 | 14 | 8 |
| *Outflow* | | | | | | | | | | | | | | | |
| *1971* | 240 | 124 | 116 | 51 | 26 | 24 | 64 | 28 | 36 | 99 | 57 | 42 | 27 | 12 | 15 |
| *1976* | 210 | 118 | 93 | 40 | 20 | 21 | 52 | 26 | 25 | 97 | 59 | 38 | 21 | 12 | 9 |
| *1981* | 233 | 133 | 100 | 49 | 25 | 24 | 51 | 29 | 22 | 108 | 64 | 44 | 25 | 14 | 11 |
| *1986* | 213 | 107 | 106 | 37 | 17 | 20 | 47 | 19 | 28 | 98 | 55 | 43 | 32 | 17 | 15 |
| *1991* | 239 | 120 | 119 | 39 | 17 | 22 | 59 | 31 | 29 | 113 | 58 | 55 | 28 | 15 | 13 |
| *1992* | 227 | 113 | 114 | 35 | 17 | 19 | 58 | 25 | 33 | 110 | 57 | 52 | 24 | 14 | 10 |
| *1993* | 216 | 113 | 103 | 32 | 20 | 11 | 49 | 20 | 30 | 106 | 56 | 51 | 28 | 17 | 11 |
| *1994* | 191 | 92 | 98 | 26 | 15 | 11 | 48 | 19 | 29 | 95 | 49 | 46 | 23 | 10 | 13 |
| *1995* | 192 | 102 | 90 | 29 | 14 | 15 | 54 | 24 | 31 | 85 | 52 | 33 | 24 | 13 | 11 |

*Source: Population Trends* 89, table 18
Office for National Statistics, Crown Copyright

*Note:*
*a* Figures in this table are derived from the *International Passenger Survey* and exclude migration between the UK and the Irish Republic. It is highly likely that they also exclude persons seeking asylum after entering the country, and other short-term visitors granted extensions of stay.

*Table J* Abortions: marital status, age and gestation (England and Wales)[a]

| | *All ages (thousands)* | | | | | *All women (thousands)* | | | | | *Gestation (weeks)* | | | | |
|---|---|---|---|---|---|---|---|---|---|---|---|---|---|---|---|
| | *All women* | *Single women* | *Married women* | *Other*[b] | *Rate*[c] | *Under 16* | *16–19* | *20–34* | *35–44* | *45 and over* | *Age not stated* | *Under 13* | *13–19* | *20 and over* | *Not stated* |
| 1971 | 94.6 | 44.3 | 41.5 | 8.7 | 8.4 | 2.30 | 18.2 | 56.0 | 15.9 | 0.45 | 1.80 | 70.4 | 20.6 | 0.85 | 2.69 |
| 1976 | 101.9 | 50.9 | 40.3 | 10.7 | 8.9 | 3.43 | 24.0 | 57.5 | 14.7 | 0.48 | 1.79 | 82.1 | 15.3 | 0.98 | 3.56 |
| 1981 | 128.6 | 70.0 | 42.4 | 16.1 | 10.6 | 3.53 | 31.4 | 74.9 | 17.6 | 0.56 | 0.56 | 108.5 | 17.4 | 1.72 | 1.02 |
| 1991 | 167.4 | 110.9 | 37.8 | 18.7 | 13.1 | 3.16 | 31.1 | 114.7 | 17.9 | 0.41 | 0.01 | 147.5 | 17.8 | 2.07 | 0.00 |
| 1993 | 157.8 | 103.8 | 35.4 | 18.7 | 12.3 | 3.08 | 25.8 | 109.7 | 18.8 | 0.49 | 0.01 | 140.4 | 15.6 | 1.84 | 0.00 |
| 1994 | 156.0 | 102.2 | 34.5 | 19.3 | 12.1 | 3.22 | 25.1 | 108.1 | 19.1 | 0.44 | 0.01 | 138.9 | 15.4 | 1.85 | 0.00 |
| 1995 | 153.1 | 101.5 | 32.7 | 18.9 | 12.7 | 3.24 | 24.7 | 105.7 | 19.1 | 0.45 | 0.00 | 136.7 | 14.6 | 1.81 | 0.00 |
| 1996 | 166.4 | 113.1 | 33.9 | 19.4 | 12.9 | 3.60 | 28.5 | 112.9 | 21.0 | 0.42 | 0.01 | 147.5 | 16.7 | 2.14 | 0.00 |

*Source: Population Trends* 89, table 17
Office for National Statistics, Crown Copyright

*Notes:*
*a* Residents only
*b* Includes divorced, widowed, separated and not stated
*c* All women per 1,000 women aged 14–49

# BIBLIOGRAPHY


Abrahams, M. (1945) *The Population of Britain: Current Trends and Future Problems*, London: Allen & Unwin.

Allen, J. and Hamnett, C. (eds) (1991) *Housing and Labour Markets: Building the Connections*, London: Unwin Hyman.

Armitage, R. (1986) 'Population projections for English local authority areas', *Population Trends*, 43, 31–9.

Ashley, J., Smith, T. and Dunnell, K. (1991) 'Deaths in Great Britain associated with the influenza epidemic of 1989/90', *Population Trends*, 65, 16–20.

Benjamin, B. (1989) *Population Statistics: A Review of UK Sources*, Aldershot: Gower.

Brooks, E. (1973) *This Crowded Kingdom: An Essay on Population Pressure in Great Britain*, London: Knight.

Chadwick, E. (1842) *Report on the Sanitary Condition of the Labouring Population of Great Britain*, edited with an introduction by M.W. Flinn (1965) Edinburgh: Edinburgh University Press.

Champion, A.G. (ed.) (1989) *Counterurbanization: The Changing Pace and Nature of Population Deconcentration*, London: Arnold.

Champion, A.G. (1993) *Population Matters: The Local Dimension*, London: Paul Chapman Publishing.

Champion, A.G. and Congdon, P.D. (1988) 'Recent trends in Greater London's population', *Population Trends* 53: 7–17.

Champion, A.G. and Fielding, A.J. (eds) (1992) *Migration Processes and Patterns, Vol. 1: Research Progress and Prospects*, London: Belhaven.

Champion, A.G., Green, A.E., Owen, D.W., Ellin, D.J. and Coombes, M.G. (1987) *Changing Places: Britain's Demographic, Economic and Social Complexion*, London: Arnold.

Champion, A.G. and Townsend, A.R. (1990) *Contemporary Britain: A Geographical Perspective*, London: Arnold.

Champion, A.G., Wong, C., Rooke, A., Dorling, D., Coombes, M.G. and Brunsdon, C. (1996) *The Population of Britain in the 1990s: A Social and Economic Atlas*, Oxford: Clarendon Press.

Charles, E. (1936) *The Menace of Under-population: A Biological Study of the Decline of Population Growth*, London: Watts & Co.

Charlton, J. (1996) 'Which areas are healthiest?', *Population Trends* 83: 17–24.

Coale, A.J. and Watkins, S.C. (1986) *The Decline of Fertility in Europe*, Princeton: Princeton University Press.

Coleman, D. and Salt, J. (1992) *The British Population: Patterns, Trends and Processes*, Oxford: Oxford University Press.

Congdon, P. and Champion, A.G. (1989) 'Recent population shifts in South-East England and their relevance to the counterurbanization debate', in M. Beheny and P. Congdon (eds) *Growth and Change in a Core Region; the Case of South-East England*, London: Pion, 106–29.

Coombes, M. and Charlton, M. (1992) 'Flows to and from London: a decade of change', in J. Stillwell, P. Rees and P. Boden (eds) *Migration Processes and Patterns, Vol. 2: Population Redistribution in the United Kingdom*, London: Belhaven, 56–77.

Dale, A. and Marsh, C. (1993) *1991 Census User's Guide*, London: HMSO.

Department of the Environment (1995) *Projections of Households in England to 2016*, London: HMSO.

Devis, T. (1984) 'Population movements measured by the NHS Central Register', *Population Trends* 36: 18–24.

Dunnell, K. (1995) 'Population review: (2) Are we healthier?', *Population Trends*, 82: 12–18.

Ehrlich, P. (1971) *The Population Bomb*, London: Pan Books.

Employment Department (1990) *Labour Market Quarterly Report*, August, Sheffield: Employment Department Skills and Enterprise Network.

Employment Gazette (1991) 'Labour force trends: the next decade', May.

Ermisch, J. (1983) *The Political Economy of Demographic Change: Causes and Implications of Population Trends in Great Britain*, London: Heinemann.

Ermisch, J. (1990) *Fewer Babies, Longer Lives*, York: Rowntree Foundation.

Fielding, A.J. (1989) 'Inter-regional migration and social change: a study of South-East England based upon data from the Longitudinal Study', *Transactions of the Institute of British Geographers*, NS 14.1: 24–36.

Filakti, H. and Fox, J. (1995) 'Differences in mortality by household tenure and by car access from the OPCS Longitudinal Study', *Population Trends* 81: 27–30.

Financial Times (1992) 'Costings are tight in an ageing market', 23 September.

Flowerdew, R. (1991) 'Classified residential area profiles and beyond', *Mapping Awareness*, 5,3: 34–9.

Fox, A.J. (1990) 'The work of the National Health Service Central Register', *Population Trends* 62: 29–32.

Fox, A.J., Jones, D. and Moser, K. (1985) 'Socio-demographic differentials on mortality 1971–81', *Population Trends* 40: 10–15.

Freeman, M. (1970) 'Not by bread alone: anthropological perspectives on optimum population', in L.R. Taylor (ed.) *The Optimum Population for Britain*, London: Academic Press, 139–49,

Gale, A.H. (1959) *Epidemic Diseases*, Harmondsworth: Penguin.

Glass, D.V. and Grebenik, E. (1954) *The Trend and Pattern of Fertility in Great Britain: A Report on the Family Census of 1946*, London: HMSO.

*Guardian* (1991a) 'Poll tax may be key to census shortfall', 23 July.

*Guardian* (1991b) 'Complacency opens door to fast moving epidemic', 23 October.

*Guardian* (1991c) 'Toll of smoking rated at 12 deaths every hour', 26 November.

*Guardian* (1993a) 'Number of lone parents doubled', 6 July.

*Guardian* (1993b) 'Lone mothers face benefits cut onslaught', 9 November.

*Guardian* (1995a) 'High risk strategy', 30 October.

*Guardian* (1995b) 'The nuclear explosion', 1 November.

Haskey, J (1989) 'Current prospects for the proportion of marriages ending in divorce', *Population Trends* 55: 34–7.

Haskey, J. (1991) 'Estimated numbers and demographic characteristics of one-parent families in Great Britain', *Population Trends* 65: 35–48.

Haskey, J. (1993a) 'Trends in the numbers of one-parent families in Great Britain', *Population Trends* 71: 26–33.

Haskey, J. (1993b) 'Lone parents and married parents with dependent children in Great Britain: a comparison of their occupation and social class profiles', *Population Trends* 72: 34–44.

HMSO (1991) *Aspects of Britain: Ethnic Minorities*, London: HMSO.

Holmes, C. (1982) 'The impact of immigration on British Society 1870–1980', in T. Barker and M. Drake (eds) *Population and Society in Britain 1850–1980*, London: Batsford, 172–202.

Howe, G.M. (1960) 'The geographical distribution of cancer mortality in Wales 1947–53', *Transactions of the Institute of British Geographers*, 28: 199–214.

Hubback, E.M. (1946) *The Population of Britain*, London: Penguin Books.

Jackson, S. and Timmins, G. (1989) 'Demographic Changes 1701–1981' in R. Pope (ed.) *Atlas of British Social and Economic History since c. 1700*, London: Routledge, 134–49.

Jones, C. (1992) 'Fertility of the over-thirties', *Population Trends* 67: 10–16.

Joshi, H. (ed.) (1989) *The Changing Population of Britain*, Oxford: Blackwell.

Kelsall, R.K. (1989) *Population in Britain in the 1990s and Beyond*, Stoke-on-Trent: Trentham Books.

Lawton, R. (1974) 'England must find room for more' *Geographical Magazine*, February: 179–84.

Lewis-Fanning, E. (1949) *Report of the Enquiry into Family Limitation and its Influence on Human Fertility During the Past Fifty Years*, London: HMSO.

McKeown, T. (1976) *The Modern Rise of Population*, London: Arnold.

Massey, D. (1984) *Spatial Divisions of Labour: Social Structures and the Geography of Production*, London: Macmillan.

Meadows, D.H., Meadows, D.L., Randers, J. and Behrens, W.W. (1972) *The Limits to Growth*, London: Pan Books.

Mitchell, B.R. and Deane, P. (1962) *Abstract of British Historical Statistics*, Cambridge: Cambridge University Press.

Mitchison, R. (1977) *British Population Change Since 1860*, London: Macmillan.

Nissel, M. (1987) *People Count: A History of the General Register Office*, London: HMSO.

Office of Population Censuses and Surveys (OPCS) (1987a) *Birth Statistics: Historical Series of Statistics from Registrations of Births in England and Wales 1837–1983*, London: HMSO.

OPCS (1987b) *Demographic Review: A Report on Population in Great Britain*, London: HMSO.

OPCS (1989) *Population and Health Monitor Series PP2, National Population Projects*, no. 2, London: HMSO.

OPCS (1990) *Mortality and Geography*, London: HMSO.

OPCS (1991a) *Census of Population 1991 (Preliminary Report for England and Wales)*, London: HMSO.

OPCS (1991b) *Deaths by Cause, Series DH2*, no. 2, London: HMSO.

OPCS (1991c) *General Household Survey: Cigarette Smoking 1972 to 1990, Series SS*, no. 3, London: HMSO.

OPCS (1991d) *Population and Health Monitor Series PP3, Subnational Population Projections*, no. 1, London: HMSO.

OPCS (1992) *1991 Census: Great Britain, National Monitor*, London: HMSO.

OPCS (1993a) *1991 Census: Report for Great Britain*, vol. 1, London: HMSO.

OPCS (1993b) *1991 Census: Topic Monitor – Limiting Long-term Illness, Great Britain*, London: HMSO.

OPCS (1993c) 'A review of 1991', *Population Trends* 71: 1–14.

OPCS (1993d) *General Household Survey 1991*, London: HMSO.

OPCS (1995) *Population and Health Monitor Series PP1, Mid-year Population Estimates*, no. 1, London: HMSO.

Owen, D. and Green, A. (1991) 'Migration differentials', paper presented to the Annual Conference of the Institute of British Geographers, Sheffield, January 1991.

Park, R.E., Burgess, E.W. and McKenzie, R.D. (1925) *The City*, Chicago: University of Chicago Press.

Paulson-Box, E.M. (1994) 'The penetration of ethnic foods into the UK diet', unpublished Ph.D. thesis, Liverpool John Moores University.

Population Panel (1973) *Report of the Population Panel*, London: HMSO.

Population Statistics Division OPCS (1980) 'Extending the electoral register canvass: a feasibility study', *Population Trends* 21: 20–25.

Ravenstein, E.G. (1885) 'The laws of migration', *Journal of the Statistical Society*, 48: 167–227.

Rowntree, J. (1990) 'Population estimates and projections', *Population Trends* 60: 33–4.

Royal Commission on Population (1949) *Report*, London: HMSO.

Scottish Office (1993) *The Scottish Diet: Report of a Working Party to the Chief Medical Officer for Scotland*, Edinburgh: The Scottish Office.

Stillwell, J., Rees, P. and Boden, P. (1991) 'Geographical patterns of migration', paper presented to the Annual Conference of the Institute of British Geographers, Sheffield, January 1991.

Stillwell, J., Rees, P. and Boden, P. (eds) (1992) *Migration Processes and Patterns, Vol. 2: Population Redistribution in the United Kingdom*, London: Belhaven.

Taylor, L.R. (ed.) (1970) *The Optimum Population for Britain*, London: Academic Press.

Thane, P. (1989) 'Old age: burden or benefit?', in H. Joshi (ed.) *The Changing Population of Britain*, Oxford: Blackwell, 56–71.

Thatcher, R. (1984) 'A review of the 1981 census of population in England and Wales', *Population Trends* 36: 5–9.

Thompson, J. (1971) 'The growth of population to the end of the century', *Social Trends* 1: 21–32.

Van den Berg, L.M., Derwent, R., Klassen, L.H., Rossi, A. and Vijverberg, C.H.T. (1982) *Urban Europe: A Study of Growth and Decline*, Oxford: Pergamon.

Warnes, A.M. (1992) 'Temporal and spatial patterns of elderly migration', in J. Stillwell, P. Rees and P. Boden (eds) *Migration Processes and Patterns, Vol. 2: Population Redistribution in the United Kingdom*, London: Belhaven, 248–70.

White, P. (1990) 'A question on ethnic group for the census: findings from the 1989 census test', *Population Trends* 59: 11–19.

Whitehead, F. (1987) 'The use of registration data for population statistics', *Population Trends* 49: 12–17.

Williams, R.M. (1972) *British Population*, London: Heinemann.

Woods, R. (1979) *Population Analysis in Geography*, London: Longman.

Woods, R. (1987) 'Approaches to the fertility transition in Victorian England', *Population Studies* 41: 283–311.

Woods, R. and Smith, C.W. (1983) 'The decline of marital fertility in the late nineteenth century: the case of England and Wales', *Population Studies* 37: 207–25.

Woods, R. and Woodward, J. (1984) *Urban Disease and Mortality in Nineteenth-century England*, London: Batsford.

Wormald, P. (1991) 'The 1991 census – a case for concern?' *Population Trends* 66: 19–21.

Wrigley, E.A. and Schofield, R.S. (1981) *The Population History of England: A Reconstruction*, London: Arnold.

Young, M. and Willmott, P. (1962) *Family and Kinship in East London*, Harmondsworth: Penguin.

# INDEX